THE ALARMIST

THE ALARMIST

FIFTY YEARS MEASURING CLIMATE CHANGE

DAVE LOWE

Victoria University of Wellington Press

VICTORIA UNIVERSITY OF
WELLINGTON
TE HERENGA WAKA

Victoria University of Wellington Press
PO Box 600, Wellington
New Zealand
vup.wgtn.ac.nz

A catalogue record is available from the
National Library of New Zealand.

ISBN 9781776564187

Printed in Singapore by Markono Print Media Pte Ltd

alarmist (pre 2020):
Someone who exaggerates a danger and
so causes needless worry or panic.

alarmist (post 2020):
Someone who justifiably raises the alarm about
a global danger to Earth's biosphere.

For Irena, Greg, Suzanne and Johanna

CONTENTS

INTRODUCTION

A southerly storm at Baring Head, New Zealand, can be a terrifying experience. The wind screams in from the Southern Ocean and races over the cliff edge with a force that numbs mind and body. The noise shrieks by at 40 metres per second like a *Count Dracula* soundtrack, a bloodcurdling whine accompanied by an eerie howling that varies in pitch by octaves. Anything not well bolted or screwed down blows away, never to be seen again; anemometers designed to measure wind speed routinely self-destruct in the gales. Huge waves pound the beach – my colleague Peter swears he saw one 12 metres high crash onto the rocks below us. I remember a storm that lasted more than seven days. Since the late 1800s many ships have wrecked in the vicinity and it's easy to see why. It's not a place for the faint-hearted, especially at night, when the lighthouse keepers worry about the ghosts of seafarers long since drowned.

Baring Head is the sampling station where I spent countless days and nights alone, making the first ever continuous baseline atmospheric carbon dioxide (CO_2) measurements in the southern hemisphere. The work was arduous and demanding and came at huge personal cost. Exhaustion and loneliness were my constant companions. It was 1972, and those and subsequent measurements at the site confirmed that humanity's impact on the atmosphere was a global phenomenon – a dreadful discovery I have lived with for fifty years.

Half a century ago, serendipity set me – a twenty-two-year-old physics graduate – on a path to becoming one of a small group who provided proof that human activities were damaging the atmosphere by dramatically altering its chemical and physical properties. Our measurements showed that atmospheric CO_2 was increasing around New Zealand as well as in the northern hemisphere. My work since then has taken me around the world, measuring minute quantities of trace gases critical to the health of the atmosphere and glimpsing climates long past through imprints left in polar ice cores. Along the way I've faced plenty of setbacks – countless failed experiments and dead ends, battles with decrepit equipment, the frustration of dealing with administrators disinterested in the science, and politicians incapable of comprehending the unimaginable consequences of the ever-increasing CO_2 in the atmosphere.

As I write, all around me in this tiny Petone home office are symbols of my lifelong journey with the atmosphere. Books, photographs, posters and scientific papers. In a simple wooden frame, a certificate: the 2007 Nobel Peace Prize. It is a testament to what I've achieved and at the same time a reminder of the things that my lifetime's work in atmospheric research has not changed. The burning of fossil fuels has continued at a terrifying pace. Atmospheric CO_2 has become the principal cause of human-induced climate change and a major driver of what is now known as a climate emergency.

The Alarmist chronicles my fifty-year journey with the atmosphere as one of elation and despair. But the atmosphere itself has a history dating back to the dawn of time, one which will continue when we are long gone. How has it changed with time and what have I seen during my own life?

Since the end of the last Ice Age about 10,000 years ago, humans have enjoyed a remarkably pleasant planet endowed with a relatively stable climate and an abundance of resources. During that period the human population has exploded from an estimated five million to around eight billion in 2021, and extraordinary advances in science, engineering and medicine have benefited most though not all of humanity. Human development over the last 300 years has been particularly remarkable: we live easier, healthier and longer lives than those endured by our ancestors. In the last twenty years the fraction of the global population living in extreme poverty has halved, almost all children living today are vaccinated, and most people have access to schooling and electricity. Is it a wonderful success story? Well, yes and no.

Rivers are polluted with toxins from agriculture and industry and wars are fought over access to clean water. Smoke from coal fires and factories have led to respiratory deaths in cities, and despite Clean Air Acts, many cities in rapidly developing and even developed countries still have toxic levels of air pollution and photo-chemical smog, leading to an estimated seven million premature deaths each year. Massive swathes of smoke from forest clearing in the Amazon and Africa are clearly visible from space and huge fires stoked by climate change are becoming more common in places as remote as Siberia and northern Alaska. In Australia catastrophic fires caused by unprecedented drought and heat raged for months in 2019 and 2020, killing an estimated one billion animals. An extreme fire season in many parts of the northern hemisphere meant that, between August and mid-October 2020, 8500 individual fires had burned almost 20,000 square kilometres, more than 4% of California's total land area and their largest wildfires in recorded history.[1] Even the vast oceans, covering more than 70% of the planet, are increasingly contaminated by household and industrial waste. In 2019 plastic waste was found in the Mariana Trench in the western Pacific Ocean at a depth of almost 11 kilometres, the deepest part of the earth's oceans.[2] In the same year Norwegian researchers made the shocking discovery that areas of the Arctic carry high loads of microplastics, more than 12,000 particles per litre of sea ice, levels that are becoming increasingly dangerous for marine life as well as birds.[3] In 2021 there is virtually no patch on our planet that does not show some trace of human activity.

Geological eras are usually measured in hundreds of millions of years, but within the space of a single century humans are driving the planet into a completely new era, one for which a new name has been conceived: the Anthropocene. For the first time in history, a single species is modifying the web of life itself, driving the natural equilibria of the Earth System into uncharted territory.

In 1960s Taranaki, when the earth's human population was only about three billion – less than half what it is today in 2021 – I'd already seen the effects of local pollution on the surrounding atmosphere. Smoke from rubbish and other fires could blot out the sun and landscape with a pungent haze in the air. But after a day of rain or strong winds the smoke was gone and the land would sparkle under bright sunshine. The atmosphere seemed to have cleansed itself – or had it?

When fossil fuels are burned, 'extra' CO_2 is released into the atmosphere, adding to 'natural' levels of CO_2 produced by natural processes like photosynthesis, where plants convert sunlight, water and CO_2 into sugars and energy. But CO_2 also drives a significant part of Earth's natural greenhouse effect: an atmospheric process where 'greenhouse gases' like CO_2 and water vapour keep our planet about 30°C warmer than it would be without the gases. This CO_2 is an integral part of the active Earth System; it's not used up, as the carbon is simply transferred from one form to another during the natural workings of the web of life and other geophysical processes. The total amount of carbon is conserved.

However, from the beginning of the first Industrial Revolution in 1750, humans began to add increasing amounts of extra CO_2 to the active Earth System by burning coal, oil and gas. I call it 'extra' because, until humans extracted it, that carbon had been locked away from the earth's active reservoirs in strata below the planet's surface.

The industrial benefits of fossil fuel combustion were obvious and far-reaching but no one is sure when the harmful impacts of adding extra CO_2 to the atmosphere were first discussed. Most climate change scientists refer to Svante Arrhenius, who published a paper in 1896 entitled 'On the Influence of Carbonic Acid in the Air upon the Temperature of the Ground'.[4] Arrhenius, a Swedish chemist, is a distant relative of Greta Thunberg – the schoolgirl who in 2018 started the 'school strikes for climate' movement and galvanised young people worldwide.

In the first part of the twentieth century no one really knew the exact effect of burning fossil fuels. Atmospheric CO_2 measurements were sparse, of uncertain quality, variable, and limited to urban areas where concentrations were likely to be higher due to local sources of CO_2. Two world wars, scores of smaller conflicts, an economic depression, the beginning of the Cold War and a host of other issues put the brakes on many areas of scientific research. Weapons development and technological research related to industrial growth were well supported, whereas geophysical and ecological research received only a fraction of the available funding. On top of this, the existing meteorological records showed only a small global temperature increase from 1900 to 1940 with, if anything, a decrease over the following thirty or forty years.

This situation changed in the 1950s. A resurgence of interest in the environment created more funding in ecological and geophysical research. Air and water pollution became obvious problems and related food contamination a public issue.

In 1953 a young American scientist, Charles David (Dave) Keeling, began a project investigating carbonates dissolved in groundwaters in California. He quickly realised he needed to know the concentration of atmospheric CO_2 where his samples were collected. But there were no useful data. Back then, estimates for background CO_2 ranged widely, from less than 150 to more than 350 parts per million (ppm). By 1956, using a laborious chemical technique on a wide variety of air samples, Keeling discovered that the background atmospheric CO_2 concentration was about 310 ppm along the Pacific coast of the US. But what were *global* concentrations, and were they increasing?

Keeling received funding to install a continuous atmospheric CO_2 analyser at a remote mountain meteorological station, Mauna Loa, located in Hawai'i, at an altitude of 3400 metres. Because this site was distant from any local sources of CO_2, it proved to be ideal for sampling air representative of the tropical Pacific.

Within three years, Keeling made two extraordinary discoveries in the northern hemisphere which completely changed our understanding of atmospheric CO_2 and shaped the way we think about climate change. First, he detected an annual cycle in CO_2 levels, with a maximum concentration in late autumn and a minimum in late spring. He realised that he'd made the first ever observations of the planet 'breathing', a completely natural process driven by the interaction of plant growth with atmospheric changes between seasons. But his second discovery was an ominous harbinger of things to come and anything but natural: the average concentration of atmospheric CO_2 was increasing inexorably year by year, the first proof that humans were conducting an unwitting and dangerous experiment on the atmosphere of our only planet.

Over the next few years, Keeling expanded his network of measurements using air samples collected on ships and aircraft. He surmised that the southern hemisphere, the 'ocean' hemisphere, would show a different atmospheric CO_2 signal, but its exact makeup was unknown. It was also suspected that the southern oceans could be a major 'sink' for CO_2. Perhaps they were absorbing a lot of the excess CO_2 derived from burning fossil fuels? To answer these critical questions, good air-sampling sites in the southern hemisphere were urgently needed.

In 1961 Keeling visited New Zealand as a possible location for a monitoring site. He met key science personnel and looked at potential sampling sites in both the North and South Islands, laying the foundation for a collaboration between his Scripps Institution of Oceanography in California and New Zealand atmospheric science teams, which has lasted for over fifty years. Dave Keeling was to become my employer, mentor and friend.

Our current geological age – the Holocene – has existed for about 11,000 years, with the atmospheric CO_2 stable at around 280 ppm for all but the last 250 years. Over the decades I've watched in disbelief as the concentration of CO_2 and other greenhouse gases, like methane, have climbed; slowly when I first started measuring them, but now, in the twenty-first century, with a growth rate that is accelerating. In 1970, when I began measuring atmospheric CO_2 at Makara, New Zealand for Dave Keeling, levels were at 321 ppm with a growth rate of 1 ppm per year. Today, in 2021, the latest atmospheric CO_2 measurements from Baring Head are more than 410 ppm with a growth rate of over 2 ppm per year.

I spent nine years working as a junior scientist measuring atmospheric CO_2. But as my research expanded, I realised that I needed to improve my understanding of the chemistry that was driving rapid atmospheric change, and so I began working in 'atmospheric chemistry'. At that time, the field was in its infancy and there was no agreement as to what the fledgling science included and what it did not. These days the area is much better defined, and engages thousands of researchers worldwide. However, even now there are still 'fuzzy edges' and good-natured arguments among scientists. A typical discussion over a beer might be, 'Does atmospheric chemistry include atmospheric CO_2? Is city air pollution a separate field? What about acid rain?'

Pragmatically, atmospheric chemistry is about solving problems related to changes in atmospheric composition and developing the best analytical and modelling tools and skills to tackle each issue. When I began studying for a PhD at an atmospheric chemistry institute in Jülich, West Germany, my research was one of twenty multidisciplinary projects that covered wide-ranging chemistry and physics as well as meteorology, oceanography, geology, volcanology and mathematical modelling. I began to learn more and more about atmospheric composition. Most of the atmosphere contains variable amounts of water vapour, with levels averaging around 1%. In dry air, 99.9% is composed of just three gases: nitrogen (78%), oxygen (21%) and argon (0.9%). Nitrogen and oxygen are vital for life on Earth but, under normal atmospheric conditions, they are chemically non-reactive. In most of the atmosphere, especially the lower part where humans live, there is no reaction between nitrogen and oxygen, and the third gas, argon, is inert.

So, how can there be any atmospheric chemistry – what's even left?

Well, literally hundreds of gases and atmospheric compounds, most of which the average person has never heard of. Incredibly, the entire field of atmospheric chemistry focuses on the characteristics of less than 0.1% of the atmosphere's composition. This tiny fraction has a profound effect on the properties of the atmosphere, providing attributes that are essential for life. They are often simply called 'trace gases' or 'trace species'. The most well-known and abundant of these is CO_2, with a concentration of around 0.03% when I started measuring it at Makara and Baring Head. In 2021, its concentration is over 0.04%.

We researchers and atmospheric chemists sounded the alarm over forty years ago. Back then, manning the station at Baring Head had been a gruelling personal project run on a shoestring. Now, things are different: there is strong support for climate change research by the New Zealand government and scientific institutes. But acknowledging consequences – of rising sea levels, record temperatures, diminishing polar sea ice, coral bleaching, increases in droughts and floods, to name only a few – is not the same as taking action. Despite huge improvements in atmospheric measurements, regular media reports and dozens of international climate change conferences, many oil, coal and gas companies are planning to increase production over the coming decade due to projected increases in demand. Our use of fossil fuels has almost doubled since 1970.[5] We know it's happening; the warnings and effects on our environment are becoming more evident with each passing year. What will it take for our species to take heed? There is so little time left.

In the 1970s when I talked about my climate change concerns, reactions were mixed and ranged from lack of interest to disbelief and even laughter. Since then I've frequently felt battered and bruised by climate deniers and have similarly suffered 'science managers' with no passion for, or understanding of, atmospheric measurements. Monitoring the health of the atmosphere is often underrated, frequently subject to budget cuts and scorned by bureaucrats who have no interest in the measurements that underpin our knowledge of climate change. Those data are often recovered at great personal cost by dedicated science teams working in remote locations. In 2021 there is widespread acceptance that humans and the planetary web of life that sustains us are threatened by a climate emergency. Significant reductions in carbon emissions must begin immediately if we are to avoid dangerous climate change. Global temperatures are already at least 1°C above pre-industrial levels, and increasing as we add more and more carbon to the atmosphere. If we continue that trend, projections show that by 2100 global temperatures will be 2 to 4°C higher than today. Already at 1°C we are witnessing unprecedented disruption caused by climate change; at 2°C many scientists predict that climate tipping points will be reached, such as the irreversible melting of polar ice sheets; but the consequences of a 4°C increase are unthinkable – there is little doubt that the civilisation we currently enjoy would largely disappear, requiring the surviving humans to adapt to extreme conditions.

To have a chance of limiting global heating to less than 2°C above pre-industrial times, global carbon emissions must halve by 2030 and reach net zero carbon by 2050. These targets were agreed to by virtually all nations at the Paris Climate Agreement in 2015 and ratified by most over the following years. But the first target is only ten years away and carbon emissions continue to increase.

Over the years I've given hundreds of presentations, public talks and interviews on climate change. I've written popular articles, published my work in scientific journals and spoken at parliament. But in a world driven by corporate self-interest and short-term politics, my words and those of hundreds of other climate scientists have not had the effect that they should. Our measurements have continued to document the accelerating rise of atmospheric CO_2 and other greenhouse gases. I've done my best calling for change – I've spent my whole life working at this.

My neck aches, my eyes are sore and my seventy-year-old body is complaining again, but I am struck by the goldmine I have discovered. I'm sitting in a library in Avalon, Lower Hutt, with boxes of files containing the New Zealand–Scripps project correspondence from the 1960s and 70s. A few months ago, I was told all of this material had been shredded in the 1990s during the transformation of the country's Department of Scientific and Industrial Research into Crown Research Institutes. But far-sighted librarian Maggie Dyer had recognised the value of these and other records and sent them to the New Zealand National Archives in Wellington. Now, years later, she has requisitioned them for me and I am spending weeks going through them and scanning letters from my friends and colleagues Dave Keeling, Arnold Bainbridge and Athol Rafter, as well as many I wrote myself all those years ago. I'm reading about the first atmospheric CO_2 measurements recorded at Makara and I'm transported back in memory, into the shoes of that young physics graduate.

PART I

1946–1972

310–326 ppm

Determination of the Buffering factor β 14/8/70

(Neglecting for a first approximation the effect of Boric acid on the equilibrium)

let total CO_2 in sea water be Σ

then Σ = $\{CO_2\}$ + $\{HCO_3^-\}$ + $\{CO_3^{2-}\}$

$$= \{CO_2\}\left(1 + \frac{k_1'}{\{H^+\}} + \frac{k_1' k_2'}{\{H^+\}^2}\right)$$

using $k_1' = \frac{\{HCO_3^-\}\{H^+\}}{\{CO_2\}}$; $k_2' = \frac{\{H^+\}\{CO_3^{2-}\}}{\{HCO_3^-\}}$

write $\{CO_2\}$ = C.

$$\therefore \; d\Sigma = C\left(-\frac{k_1'}{\{H^+\}^2} - \frac{2 k_1' k_2'}{\{H^+\}^3}\right)d\{H^+\} + \left(1 + \frac{k_1'}{\{H^+\}} + \frac{k_1' k_2'}{\{H^+\}^2}\right)$$

$$d\Sigma = -\frac{C}{\{H^+\}}F_2 \, d\{H^+\} + F_0 \, dC$$

$F_0 = k_1' k_2' / H^2$

where $F_0 = 1 + \frac{k_1'}{\{H^+\}} + \frac{k_1' k_2'}{\{H^+\}^2}$ change F_0 & F_1 around

$F_2 = \frac{k_1'}{\{H^+\}} + 2\frac{k_1' k_2'}{\{H^+\}^2}$

$F_3 = \frac{k_1'}{\{H^+\}} + 4\frac{k_1' k_2'}{\{H^+\}^2}$

$$\frac{d\Sigma}{\Sigma} = \frac{-C \, F_2 \, d\{H^+\}}{\{H^+\} \, C \, F_0} + \frac{F_0 \, dC}{C \, F_0} \quad \cdots\cdots \; ①$$

CHAPTER 1

A SURFING SALVATION

Surfing is about the waves, the wind and the atmosphere. As a teenager, I would sit in silence on my surfboard a hundred metres or so out from the beach, above a reef. I'd look out to sea, feeling easy swells pass beneath me as I waited for the next set of larger, surfable waves. On the horizon the atmosphere seemed to disappear into the ocean in an enormous arc, a hemisphere of blues, greens, whites and greys extending as far as I could see. Wavy lines in the distance would signal the arrival of the next set of waves and I'd swing round, paddling frantically as a mountain of glassy water moved towards me. Flashing across the face of a living column of water, watching the sky and sea rotate around you – often I'd scream out loud with the thrill of it. No matter how good a surfer you are, there will always be a wave that will defeat you, leaving you disoriented and buried under tonnes of swirling seawater. When I look back, I see my life as a series of tipping points, times when something or someone sent my life along a different path. One of those points was discovering surfing.

As a young surfer and ever since, I've tuned in to the atmosphere. Watching its changing moods is a rewarding and spiritual experience. Every day it puts on a different show, high wispy cirrus clouds of ice crystals change into dense, iron-grey storm clouds with wind and rain. Then after the rain has passed, rhythms of dancing light and an intoxicating smell hang in the air. Bright sunshine bakes vapours from the earth into the atmosphere. In the surf zone, light blends with waves and the smell of seaweed. All your senses work together as your mind and being harmonise with the atmosphere.

*

I grew up in rural Taranaki. Like many young people raised in the countryside, I developed an early connection with the ocean, rivers and land. It was intuitive because we were outside most of the time in the sun and rain, swimming and playing and watching changes in the atmosphere above. I didn't know it then, but the same feelings are universal. Indigenous peoples all over the planet relate to what might be called an 'Earth mother'. The Māori Earth mother, Papatūānuku, gives birth to all things, including people. We share an innate sense that the oceans, air, land and all living things are somehow joined in a web of life.

New Plymouth was a wonderful place to grow up. It was big enough that, as a teenager, you could meet lots of friends. But it was small enough that a short ride on a bike could take you to native bush around pristine lakes and rivers and the wonderful environment that encircled the town. The black-sand surf beaches and offshore reefs to the north and west gave way to the bush and farms inland, with the almost perfect volcanic cone of Mt Taranaki dominating the landscape. Back then, any notion of global air and water pollution felt remote. New Zealand's population was only 2.4 million, and to many New Plymouth's environment appeared pristine.

But was it? My surfing fraternity and others had already noticed that stored sewage was often dumped into the sea during southerly storms. If the wind forecast was wrong, raw sewage washed up on New Plymouth's beaches and floated alongside us in the waves. This shocked and disgusted me as a teenager who was beginning to appreciate the environment and the remarkable role of the oceans in shaping our planet.

I'd seen streets and roadsides littered with rubbish that people simply threw out of car windows. Precious native bush was still being logged, leaving massive scars on the land and creating steep erosion-prone hillsides. Smoke from scrub and bush burning often turned the sky grey. I was outraged. Yet in the New Zealand of sixty years ago, there was no perception that these actions were actually damaging an environment that people would want to enjoy. Pollution was not talked about and certainly not widely understood. 'Out of sight, out of mind' was a common attitude back then. The problem is, garbage is never out of sight for long. Waste disposed of in one location has a habit of turning up elsewhere. In New Zealand and many other countries, old coastal landfills are being breached by rising sea levels caused by climate change. Frequent storm events are beginning to expose dangerous toxic wastes, like asbestos and heavy metals, heedlessly disposed of decades earlier.

As a teenager I seethed with a quiet anger. You didn't need complicated mathematics or physics to understand the idea of a finite earth. If you kept on filling the air, oceans and land with wastes while exploiting non-renewable resources at an increasing rate, you were going to have both pollution and resource problems. It was obvious, even to a seventeen-year-old, that unchecked global population growth with no restraint on resource consumption and emissions was going to lead to the collapse of the ecosystems that our lives and livelihoods depended on. If you continue to fill a finite space like the atmosphere with contamination, at some stage the damage will be obvious. I had always perceived this as a matter of 'when' rather than 'if'.

Since the age of twelve I'd been attending New Plymouth Boys' High School, a hellish experience that imprinted itself forever on my psyche. Somehow, I'd started at the school about a year younger than anyone else; I was small, and was the youngest of over a thousand students at the school. From day one I was systematically bullied by bigger kids. I'd arrive in the morning on a country school bus and they'd be lurking near the school gates. First, they'd rip my bag away and tip my books over the ground. As I stumbled around trying to pick up the mess, they'd dart around me tearing at my clothes and punching my head. Often a sniggering group of boys looked on.

I wasn't the only boy targeted; other small boys were hurt and humiliated too, especially in the boarding houses. These double-storeyed dormitories were straight out of a Dickensian nightmare.

25

Here, following a bizarre parody of nineteenth-century British boarding school life, senior boys had 'fags' – young boys who were enslaved, exploited and routinely punished for random 'crimes'. Bullying at the school in those days was a noxious routine, ignored by teachers who could not have been ignorant of the inflicted terror. In my first three years of misery at the school, never once did a teacher show any kindness or intervene in any way.

Relief came each day when I scrambled into the safety of the bus for the hour's ride back to the tiny country flat where I lived with my parents and younger brother, Steve. We had just three rooms: two bedrooms, a combined kitchen/dining/laundry area, and an outhouse toilet-bathroom that had ice on the inside of the windows during winter and no fixed door. It was a dilapidated but intensely happy home, a haven tucked away at the end of an old World War Two airport with grass runways, about 15 kilometres from New Plymouth. On the other side of a small access road was a large transit camp full of displaced and homeless people with lots of children.

Every day after school there would be at least twenty kids aged from about five to fourteen playing soccer, cricket, hide and seek, bullrush – as well as games we made up, some of which were really dangerous. In one of our pretend war games we used box-wood torn out of packing crates to make razor-sharp arrows which were fired at rival 'armies', along with stones launched by catapults made out of old car tyre inner tubes. I vividly remember the whine that stones and arrows made as they ricocheted off the concrete blocks we hid behind when our particular army was under siege. The blocks were all that remained of the old mess huts and barracks that used to house the New Zealand Air Force. By today's standards, most of the families I knew – especially the people at the transit camp – were barely above the poverty line. About once a month, Dad would buy a bottle of lemonade which he split into four equal portions and, as a family, we would sit on the front step of the flat enjoying the rare treat. Did I feel poor? Absolutely not.

Life in the country was the antithesis of the dreadful experiences at high school. Away from high school, my life was an idyllic mix of playing with kids from the transit camp and local farms, and learning about electronics from my dad. He was a radio and radar engineer working at the old New Plymouth airport on communications and distance-measuring equipment for aircraft. He'd worked for twenty years at various airports in East Africa as well as spending time as a radio operator on cargo ships in the Atlantic.

Dad was passionate about his hobby as a 'ham' or radio amateur, and used home-built radio equipment in our flat to contact other radio amateurs throughout New Zealand and the world. Through him, I developed a fascination with electronics and built simple valve, tube and transistor amplifiers as well as primitive radio sets. I also built a listening device with a microphone which I set up in our parents' room behind their bed. Dad soon discovered this and – for some reason I didn't then understand – quickly removed it!

Mum was a shorthand typist with a wonderful sense of humour and a great attitude to life. She was South African and spoke fluent Afrikaans, and I think that my interest in languages later in life came from her. Dad used to shout and pace up and down when he was angry, but I don't think Steve and I ever heard a cross word from Mum.

At school I learned little in the classrooms where I was routinely kicked. Each day became a matter of survival where I desperately tried to avoid the school's most notorious bullies. Some played for the school's top rugby or cricket teams, as if this entitled them to a free run at bullying smaller boys. What happened to them in later life? I heard that at least one particularly nasty lout, Freddy, was jailed for assault. What a stupid and useless waste it all was. With enlightened behaviour from the teachers and the headmaster, dozens of boys could have been spared the misery which scarred some of them for life.

Mum and Dad were supportive and must have been concerned about the dejected boy returning home from high school each day. I suspect they never really appreciated what was happening to me. What they saw was the mud-covered happy child playing with the local kids after school. Looking back, I realise how precious those hours playing with the country kids must have been to me – they will never know how much I owe them.

After three years of misery and failure at high school I'd had enough. I was fifteen, the legal age for leaving school, and left for an entry-level job at the New Plymouth telephone exchange. My role there was making cups of tea for the senior technicians and cleaning grease off mechanical telephone equipment. It was filthy, boring work using dangerous solvents that left my hands raw and chafed, with grease packed under my fingernails. But no one picked on me.

I started meeting other young people, including the first to take up surfing on the Taranaki coast. There was Leftie, who lived in a clapped-out van with his surfboard and not much else. When he came out of the waves, without fail he would get stuck into banana sandwiches; that seemed to be all he survived on. And there was Rich, who was some sort of accountant able to get off work whenever the surf was running. He would sometimes ramble on about numbers, stop in mid-sentence, stare out to sea, then scramble into his suit to go back to work.

Surfing was a time of anticipation and elation. A bright blue sky and blazing sun overhead. Black iron sand burning your feet as you ran towards the sea, surfboard under your arm. Spray on your body as you dived into the waves clutching your surfboard, clean seawater flicking you around yet holding you in a fluid embrace. A billion or two years ago we came from the sea, didn't we?

It was during this time that I began to bond with an inspiring primary school teacher, Ray Jackson – the father of one of my best friends, Con. I could open up to Ray and talk to him in depth about my feelings in a way that seemed easy and natural. Ray always listened to me carefully and was never judgemental.

I remember feeling surprised when he asked me what I thought about books and maybe I'd like to look at a few. I'd read a lot of books when I was younger, but virtually gave up during the terrible years at high school. At the telephone exchange we read comics and magazines rather than books. I resisted Ray's suggestion at first. However, when he nudged me in the direction of the city library, I was amazed to discover books on surfing in Hawai'i and books about the environment, including on weather systems and the formation of ocean waves. Soon I found another world through the pages of library books. One I'll never forget was Rachel Carson's *Silent Spring* (1962), a sobering account of the poisoning of the biosphere with DDT and other chemicals aimed at pest, disease and weed control. Learning and learning how to learn from the books in the library was mind-blowing. I discovered books on science, mathematics, engineering and electronics, among others, and vividly remember my excitement as I became more and more aware of the science behind the physical forces shaping our planet.

Through surfing I'd already developed an almost primal feeling for ocean waves. I'd discovered from my reading that these were often generated by storm events thousands of kilometres away as energy passed from the atmosphere into the sea. Soon I'd gone through dozens of books in the New Plymouth library – the entire science section – and was rapidly reaching the limit of what I could learn there. What was my next move?

To study physical science, I would have to go to university. But I was a high-school drop-out with only a basic school qualification. I'd barely scraped through School Certificate. In a provincial New Zealand town in the 1960s, getting to grips with subjects like physics, oceanography or atmospheric science seemed a far-fetched idea. Accounting or agricultural studies, perhaps, but fundamental science? How could that be useful, let alone provide a living? After discussion with Ray and my parents it was clear I would have to go back to school for another year. Without the University Entrance certificate, I could not enrol at a university.

My workmates at the telephone exchange shook their heads when I quit to go back to school. My parents had little money, and I had to buy a school uniform and sell my motorbike to help with the family finances. I went back to the same awful school but this time it was different: I was single-minded about getting University Entrance.

When I re-enrolled in academic classes at the same school in 1964 after a year's absence, I was treated as a 'new boy'. However, I knew exactly what to expect. When told to enrol in the school military battalion, I deliberately did not. On military drill days I skipped school and went surfing instead. I did not exist in the military battalion records and was never caught, although teachers sometimes patrolled the surf beaches looking for truant boys.

In the academic school I enrolled in a mixture of science subjects, pure and applied mathematics as well as English. It was a huge challenge because my previous experiences at the school had left me poorly equipped to deal with the advanced subjects I had chosen. For the first month I struggled to keep up, but I was highly motivated and got on top of the work. My teachers were somewhat diffident about having a boy with 'broken schooling' in their classes, but I made it clear that I was there to learn. The other boys, in what turned out to be the top science class, were mature and also wanted to learn. There was none of the bullying and mindlessness I'd previously experienced. Our science textbooks were based on inquiry and laboratory experiments rather than rote learning, and encouraged students to appreciate the elegant physical principles driving the phenomena we see in everyday life. Physics involved practical calculations and experiments on sound, water and radio waves, heat, light and radiation, as well as concepts behind launching rockets and spacecraft into Earth's orbit and beyond. I topped the school in physics that year.

Chemistry caught my imagination and I was soon developing ideas at home to make rocket fuels and explosives. This was initially done in my bedroom and the garage of my parents' home. After one explosion filled the house with smoke, I was told to take the experiments elsewhere. Ray Jackson's son, Con, a willing participant in the tests I'd designed, helped me move the equipment to the back garden of their house. After one particularly fun afternoon in which a small rocket ran amok, we noticed Ray's cucumbers were wilting. The next day it was obvious they had succumbed. Ray, always sanguine, suggested we should learn from the experience. I remember helping to replant the garden and buying him some apology tinned tomatoes from the local dairy.

Other chemistry experiments were potentially more dangerous. I developed a method for making hydrogen gas by dropping thin aluminium milk bottle tops into a concentrated caustic soda solution in the bottom of a quart beer bottle. The gas was collected in rubber balloons to which I tied labels marked 'Please contact Dave Lowe phone 88–390'. These were launched to try and track how far they flew. One day my brother Steve was helping me when, without warning, he decided to light the hydrogen coming out of a beer bottle. Momentarily his head was enveloped in a red flash of flame and, when he turned to me, he had no eyebrows and was deaf for a week. Mum, gentle but assertive, suggested that we might want to move on to other less dangerous experiments.

Being a single sex school, Boys' High provided no opportunities to meet girls, let alone learn to get on with them. But I had a driver's licence, and my parents loaned me their Morris Minor to take girls out to local dances, organised by churches and other groups, in halls around New Plymouth, Stratford and Waitara. Having 'a pash', as we called it, fumbling around in the cramped back seat of the car, was a fun and uncomplicated learning experience. I was lucky to be friends with a group of girls and boys who got on really well. Conversations included who had pashed who the week before, and the latest Beatles or Rolling Stones number.

The final year at high school was remarkable in many ways. When I talked to some of the younger teachers at the school, I was astonished to find that they also felt intimidated and repressed by some of the management issues at the school. The school's headmaster kept an iron grip on discipline. The impression of most of the boys was that some of the older staff revelled in the opportunity to belt a boy with a cane. One particularly sadistic older Latin and maths teacher used a piece of machine-belting to administer punishment and routinely drew blood. In a concerted action, the younger teaching staff tried to do something about corporal punishment and failed. Widespread bullying continued well after I left and was often reported by the local press. Con and I left the school over fifty years ago and, despite repeated invitations to join the Old Boys Association', both of us cut our ties with the place forever.

When I enrolled at Victoria University of Wellington, I had no inkling of the twists and turns that lay ahead. I just knew that if I was to focus on the environment, I had to study physics, chemistry and mathematics.

As capital cities go Wellington is small, but to a country boy it was huge, daunting, dirty and depressing. I felt alone. I knew no one at the university, or anyone who lived in Wellington, and I was filled with self-doubt.

Student life at Victoria was compelling, to say the least. Student parties were usually out of control, noisy and quite dangerous. I got drunk on many occasions, threw up and grew up. After one all-night party I remember waking up hanging over a bush in someone's front garden, a disgusting pile of mature vomit below me. 'Do you really want to do this?' asked a rational voice in my head as I gradually came to with every part of my body aching and my teeth itching. I continued to enjoy student parties after that, but getting blind drunk had definitely lost its attractions.

Unlike many of the other students who were funded by wealthier parents, I had to pay all my living expenses as well as buy my own university textbooks and pay other fees. At the end of each university year my savings would be cleaned out, and I would work at various holiday jobs to earn money for the following year. One year I drove a truck around Wellington delivering soft drinks to small shops in the suburbs; I worked as an electrician on a car assembly line in Lower Hutt, and also on the Wellington wharves unloading sacks of mail and other goods. At a small dairy factory in Taranaki I helped make cheddar cheese in huge vats as well as working on sheep and dairy farms. My experiences with some colourful workmates certainly helped me put things in perspective. I felt more in tune with the working class and 'down and outers' than I did with some of the snobbish students from wealthy families in my science classes at Victoria.

In Wellington one of my close friends was Allan James, a music student and guitar player. He helped me buy my first guitar and taught me how to play it. This set the scene for me playing electric guitar for a student rock band. The band was terrible; I can't remember whether we even had a name. But the people we played for didn't seem to care. The band members were mostly from the chemistry department and we used to practise in one of the lecture theatres. I played rhythm using a homemade electric twelve-string guitar and was the lead singer. Our lead guitarist was tone deaf, which was a bit of an issue because he couldn't read music either. But he had a very good memory and learned all the notes by heart. Occasionally he would play the wrong part of a tune and, without an ear for the music, had no idea anything was wrong. I'd have to quickly try to mask the altered tune he was playing.

Being in a band introduced me to a lot of people that I otherwise would not have met. I was studying physics, chemistry, pure and applied mathematics – the so-called 'hard science' subjects. There were virtually no women in those classes. But being in the band and playing at folk clubs in Wellington meant I met a lot of young women. I also went to a lot of parties, where I met people with thoughts and opinions different from the other science students.

At one of these parties I met Rosemary Sutton, a nineteen-year-old education and psychology student. By coincidence she also came from New Plymouth, and although I had never met her there, we had several friends in common. She was immersed in her studies and fascinated me with her knowledge of psychology. I remember her charming me with a story about a psychologist who raised an infant chimpanzee with his own baby son, noting their behavioural differences over several months.

Rosemary and I were immediately attracted to each other. After a very short romance, like many other students, we got married at a small ceremony in New Plymouth followed by a reception in a coffee bar. It was all very low-key, but we were happy and continued with our university studies. Fifty years ago, everyone seemed to marry early – living together while unmarried was frowned on and it was much less complicated just to follow the norm, especially as both our sets of parents were regular churchgoers. In hindsight, Rosemary and I were way too young to understand the commitments involved in marriage. We were still growing up and neither of us knew what we really wanted out of life. We had virtually no money and no idea where we were headed. But my friends joked that at least we had my guitar and surfboard. Our love was very physical – we were close to each other and committed to the relationship. We both had very heavy university workloads which kept us in our dingy flat a lot of the time, and depending on study timetables we went out to parties with students and friends around Wellington.

It was in my physics honours year that I knew I was on track to being a research scientist. As well as the coursework, there were three large research projects where we had to devise experiments and literally find something new. One of my projects was with Paul Callaghan, who later became *Sir* Paul Callaghan and New Zealander of the Year for his groundbreaking research and contributions to society. In our experiment we had to crawl 100 metres underground along the length of a sewerage pipe access-tunnel underneath the chemistry department, setting up a helium neon laser and a series of Fresnel lenses. The idea was to try and distinguish between so called ballistic and non-ballistic photons emitted by the laser. I took a series of images on photographic plates behind the lenses and found huge interference patterns showing that all of the photons from the laser were wavelike. I remember the supervisor being incredibly disappointed by this result. He'd wanted us to find a mix of different kinds of photons. For the first time, I was aware that a scientist needs to avoid bias and to be totally objective.

I graduated at the end of 1969 at age twenty-two. On advice from one of my lecturers I applied for a job as a junior scientist at the Institute of Nuclear Sciences (INS), a division of the New Zealand Department of Scientific and Industrial Research (DSIR) based in Gracefield, a suburb of Lower Hutt, north of Wellington. I was interviewed by Athol Rafter, the director of INS, a charismatic man and a top international scientist. He had a great sense of humour, was a real judge of character, sought the best from his staff and was always interested in what they had achieved. I liked him immediately. His field was the application of both stable and radioactive isotopes to the study of the atmosphere and oceans.

At the interview he told me that atmospheric CO_2 from burning fossil fuels was probably altering the earth's climate. This was my first introduction to the calamity that was to come, and I was both shocked and amazed. That such apparently small changes in an atmospheric trace gas could cause a significant global effect seemed both unbelievable and horrifying, though just how significant we didn't yet know. Athol told me that an American group named Scripps was just beginning to set up atmospheric CO_2 measurement systems in New Zealand, using the INS as a base. I spent a wonderful morning with Athol as he showed me around the institute and introduced me to several scientists working on environmental projects involving naturally occurring stable and radioactive isotopes. He seemed taken with my background, questions and enthusiasm. He offered me a job on the spot and we agreed that I would start early in 1970. My journey with the atmosphere was about to begin.

CHAPTER 2

THE ATMOSPHERE CALLS ME

It was a weird feeling to set off on my first day of work as a research scientist. I'd planned my life around this moment, but doubts flooded through me as I arrived at the institute to report for work. The main site of the INS, where the administration section was located, was a group of brand-new architecturally designed buildings with a commanding view of the Hutt Valley and across the harbour to Wellington. My mood improved at the prospect of working there.

But I was told my first assignment would be in 'Shed 2', working on a hydrology project with a Scottish man named Claude Taylor. I'd not realised that, as well as the main site, the INS comprised a series of laboratories in old wooden sheds. These were the original site of the institute and had served as Athol Rafter's own research area before the new buildings were finished. The sheds had been built in World War Two by the US army to store military equipment used to retake the Pacific Islands occupied by the Japanese.

Shed 2 was a ramshackle building in a rather run-down industrial region of Gracefield. The entrance was through a rusty barbed-wire fence, down an alleyway, past wrecked cars and rusty oil drums. The main doorway into the INS labs was at the back of the building through a rickety lean-to, past a small block containing a cobalt-60 radioactive source and a rusty shipping container. Many of the windows were cracked or broken, and there was dirt and rubbish piled up around the walls.

Inside Shed 2 I was introduced to Claude, who ran the tritium hydrology laboratory, and shown into a dark and dingy cubicle. My heart sank. The main laboratory had a large glass vacuum line with a furnace, an electrolysis plant and a series of shelves for holding bottled water samples. But what caught my attention was the floor. As I walked on the cracked linoleum, I could see movement – a kind of silvery shimmer that seemed to come from the floorboards underneath.

'What's that?' I asked Claude.

'Don't worry about that,' he replied. 'It's only mercury. When we have a break in the vacuum line, the mercury from the pumps and manometer flies everywhere and winds up on the floor.'

I remembered from my studies that mercury was extremely toxic, even in small amounts, and accumulated in the body.

'Sometimes we put sulphur powder down to try and reduce the vapour released,' he added, 'but usually we don't bother. I wouldn't worry if I were you – I've been here ten years and have survived.'

There were ten INS laboratories in Shed 2, and later I discovered there was mercury on the floor of every single one, not to mention asbestos and rats in the roof. One of the scientists running the lab opposite Claude's was an alcoholic, and slept on the mercury-ridden floor during lunch breaks. He was barely coherent a lot of the time. I had visions of nineteenth-century factory conditions. What had I got myself into?

The work with Claude involved using measurements of tritium, a naturally occurring radioactive isotope of hydrogen, to track the movement of rain and groundwater through aquifers underground. It was fascinating and I was very curious about the science. However, Claude soon made it clear that, as far as he was concerned, I was there to work for him in a technical role only. My job was to run the vacuum line and electrolysis plant, enriching samples of water so that natural tritium levels could be detected using a counter designed to measure the radioactive decay of tritium. The water samples were collected from rivers, springs and aquifers all over New Zealand as well as the Pacific Islands. It was obvious Claude had no intention of spending much time discussing the science. He seemed to have no interest whatsoever in other people, let alone a fresh young graduate in his first job. Although Claude could be pleasant, he was often dour and taciturn, and would ignore me. I remember he once ended a phone call by abruptly hanging up on me.

Being stuck with Claude in the depressing Shed 2 started to get me down. The technical work became repetitive and boring; I felt trapped. I heard later his wife was German and desperately unhappy living in New Zealand. Many of the other scientists had written off Claude as weird, and sadly he had few friends at the institute. He certainly had no patience with my endless questions. Rosemary tried to console me, but she was finishing a very demanding final year at university and had little patience for my moping. She was studying from our Wellington flat while I spent weekdays at the INS in Lower Hutt. I'd often wind up back home late at night, exhausted from the commute and a hard, depressing day with Claude, to find her in bed asleep. Even though we had only been married a few months, we were already on different trajectories. I felt alone.

Despite the surroundings and the drudgery in Claude's lab, I enjoyed meeting the other scientists and technicians in the complex. They were a diverse bunch from different countries and backgrounds, and made me feel welcome. I enjoyed getting to know them at regular tea and lunch breaks in a communal cafeteria, while Claude preferred to eat lunch in his lab.

All of the projects at Shed 2 involved using stable and radioactive isotopes to study the environment. Isotopes are variants of the same chemical element, with different numbers of protons and neutrons in their nuclei. As a consequence, their masses are slightly different and changes in carbon isotopic ratios, for example, provide insight into natural processes that can't be obtained in any other way. One of the main projects was radiocarbon dating. Radiocarbon, or carbon-14, is a naturally occurring isotope of carbon produced in the upper atmosphere when high-energy cosmic rays bombard air. The carbon-14 technique is widely used in archaeology to determine the ages of artifacts as well as human and animal remains. It was used to date the Turin shroud, for example, settling a centuries-old controversy about the last garment supposed to have been worn by Jesus Christ.

Although the Shed 2 labs processed a lot of archaeological samples, Athol Rafter had taken the technique a step further by using it to trace the movement of carbon in the atmosphere and oceans. In the late 1950s he had shown that the carbon-14 content of atmospheric CO_2 in the southern hemisphere was being diluted by CO_2 from fossil fuels, which contained no carbon-14. It was an extraordinary piece of scientific detective work, proving that the atmosphere's naturally occurring carbon dioxide was being added to by emissions from burning coal, oil and gas.

At the end of Shed 2, about 10 metres from Claude's lab, there was a smaller lab with a handwritten sign taped to the wall outside:

Scripps Institution of Oceanography
University of California San Diego
La Jolla California
ATMOSPHERIC TRACE GAS LAB

This inconspicuous room would mark the beginning of a joint US–New Zealand programme which has endured for over fifty years, providing vital measurements of greenhouse gases for international climate science. It was in this poky lab that the equipment was designed and assembled for the first continuous atmospheric CO_2 measurements in the mid-latitudes of the southern hemisphere. I'd met the three Americans involved with the programme on my first day at Shed 2: Arnold Bainbridge, the group leader; Skip Price, a Californian technician with a background in anthropology; and Bob Williams, an ex-US Army soldier employed as a student researcher.

Their equipment had arrived from California in a shipping container stored amongst the debris behind Shed 2. From the beginning I was fascinated with their programme and talked to them about it during tea and lunch breaks. This was the first time I'd talked to Americans in any depth, and I was fascinated by descriptions of life in the US as well as the programme they were working on.

After two months in Claude's lab I decided this work was not for me. The reasons were partly personal: Claude was so abrupt and indifferent to my questions, needs and interests. There had to be a better science job than this. After discussing the situation with Rosemary, I decided to go and see Athol Rafter and resign. In a trembling voice I stood in his office and told him I wanted to leave. He looked at me.

'You can't do that. You're on one of our most valuable science projects,' he said, shaking his head. I felt intimidated, and wondered if I had locked myself into some inflexible contract. Athol seemed understanding and kind, but I could tell he was annoyed.

'Arnold Bainbridge tells me you've been talking to him. If you're interested, he's keen to have you working with his group on atmospheric CO_2.'

'Wow, yes,' I stammered.

'The snag is that Claude will need you to continue with him until we find a replacement. Hang in there for a bit.'

I nodded and stumbled my way out of the office.

That night I had celebration beers in Wellington with Allan James, my musician friend. If I could work with the Americans on their project, I'd have no problems putting up with the depressing environment at Shed 2.

The next day, when I reported for work complete with a hangover, Claude seemed to be constantly on the verge of saying something. Obviously Athol had phoned him and given him a grilling. Fifty years ago, science graduates – especially physics majors like me – were hard to find, and Athol clearly wanted to keep me at the institute.

Was I just a tell-tale? Hell no! I knew I had every right to stick up for myself. Claude's attitude had sapped my morale and enthusiasm for environmental science, and I found out later he had upset many others as well. In any case, I had not told Athol directly that I had issues with Claude – I'd only told him that I wanted out. Unsurprisingly, my working relationship with Claude deteriorated even further over the next couple of months. The tedious work in the tritium lab continued but reduced as I became more involved with the American CO_2 programme. Looking back, nothing had prepared me for the stark reality of working in a real-life situation involving problem people and unpleasant conditions. To survive I had to put my expectations to one side and learn to adapt, while making sure there was no compromise to the integrity of the science.

Oddly, I think I had high school and holiday jobs to thank for the survival and people skills that helped me through those first few months. Many of the students I'd graduated with had never had to work in hard physical jobs to make ends meet. Perhaps their more sheltered upbringings meant they struggled to adapt to a world outside the university. I was lucky: I understood that life, like my initial experiences in Shed 2, was warts and all.

PART II

1972–1975

326–330 ppm

Exchange of CO_2 between atmosphere and ocean surface

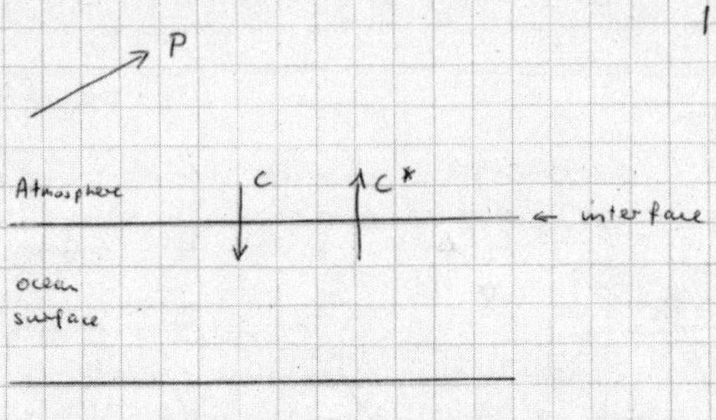

C — CO_2 in moles/mole in atmosphere

C^* — CO_2 in moles/mole in ocean surface

P — Production rate of CO_2 from burning of fossil fuel. (in atm

Na — No of moles in atmosphere ie no /atmos.

Assuming that rate one way thro interface same as rate the oth
way can write.

$$Na \frac{dC}{dt} = P Na - K A_0 (C - C^*)$$

or

$$\dot{C} = P - (C - C^*)/\tau$$

where $\tau = \frac{Na}{K A_0} = \frac{A_c \times \overline{M}}{K \times A_0} = \frac{35.5 \times 10^4}{K \times .71}$

$\overline{M}$ — No of moles above square meter earth.

A_c/A_0 — .71 ratio surface to ocean.

K —

get τ of the order of 5 years. which checks with C^{14}

$$P - \dot{C} = \frac{C - C^*}{\tau} \quad \text{p.p.m./year}$$

$$(P - \dot{C}) Na = \left(\frac{C - C^*}{\tau}\right) Na \quad \text{moles } CO_2/\text{year} \\ \text{into the sea.}$$

CHAPTER 3

MAKARA

Working with Arnold Bainbridge was the antithesis of working with Claude. For the first time I was challenged by someone who had an in-depth knowledge of both oceanographic and atmospheric science and experience in taking some of the critical measurements needed to understand human impacts on the planet. He was a brilliant scientist as well as a very practical, hands-on person with analytical equipment. He had originally worked with Athol in New Zealand in the late 1950s, before moving permanently to the US, where he'd worked at several science institutes before joining Scripps in 1966. He was about forty when I met him, a chain-smoker, very overweight, and to me had a strong American accent. We worked on a lot of electronic equipment in the lab, and he used to tell me to 'sodder' equipment rather than 'solder' using a soldering iron.

Arnold confirmed what I had already learned from Athol Rafter about the probable effects of increasing atmospheric CO_2 on Earth's climate. He also told me that in the mid-1960s a group of concerned scientists from Scripps, whom I later found included Dave Keeling and Roger Revelle (the director), had delivered a report to the President Lyndon B. Johnson's Science Advisory Committee on the risks of rising atmospheric 'carbon pollution'. That report shows that, more than sixty years ago today, these scientists already had a strong understanding of how increasing atmospheric CO_2 could have widespread effects on not only the earth's climate but sea-level rise and ocean acidification. The warnings in the text of their report were clear:

45

Through his worldwide industrial civilization, Man is unwittingly conducting a vast geophysical experiment. Within a few generations he is burning the fossil fuels that slowly accumulated in the earth over the past 500 million years . . . The climatic changes that may be produced by the increased CO_2 content could be deleterious from the point of view of human beings. The possibilities of deliberately bringing about countervailing climatic changes therefore need to be thoroughly explored.[6]

At the time, the president was preoccupied with the Vietnam War, which was going badly, as well as with serious civil rights issues in Mississippi. The evidence presented in the report was based on a series of atmospheric CO_2 measurements at Mauna Loa, a mountaintop in Hawai'i, started in the late 1950s and run by a scientist from Scripps, Dave Keeling. These showed that the concentration of the gas was rising in the atmosphere due to human combustion of fossil fuels. By the time the report was handed to the president, the atmospheric CO_2 concentration at Mauna Loa was about 317 ppm, and had been increasing relentlessly every year since Keeling first measured it in 1957. The concentration there today is now well above 410 ppm. At the time of the report, Keeling was concerned about the increase; but even he, who became the world's leading authority on atmospheric CO_2, would never have imagined the scale of the problem we face now. If President Johnson and successive administrations had acted on the advice in this report last century, it's probable that we would not be facing a climate emergency this century. The Scripps report wasn't buried, Arnold told me, 'it's just that things that are clear and important to scientists don't necessarily impress politicians!' How clairvoyant that was.

Arnold was the first to tell me the dreadful implications for the oceans caused by increasing CO_2 and temperature: that the acidity of seawater would rise as a result of excess CO_2 being absorbed in the ocean surface. A more acidic ocean would compromise the way that marine organisms formed shells, which would be exacerbated by rising ocean temperatures leading to mass coral bleaching events. At this stage very little was known about how atmospheric CO_2 was absorbed by the oceans. I was fascinated, and resolved to learn more about the process.

For the first time since leaving university, I was being pushed intellectually and scientifically. My endless questions were welcomed and answered in detail by Arnold, and he also pointed me towards the very limited amount of scientific literature available on atmospheric CO_2, most of it published by Dave Keeling and co-authors during the 1960s. At this stage, although Keeling was the world's foremost expert on atmospheric CO_2 and it was his programme I was working on in New Zealand, Arnold never ever mentioned him by name. I read Keeling's papers carefully but, with no information to the contrary, assumed that the atmospheric CO_2 programme was being run by Arnold. From him I learned areas of science that were new to me, including atmospheric composition, ocean carbonate chemistry, mathematical modelling and time-series data analysis. But at the same time, through him, I had my first introduction to the unwillingness of politicians to become involved in serious scientific issues.

After I'd been at the INS for four months, Rosemary and I decided to move out to a tiny flat in the Hutt Valley, where she could begin teacher training at an outpost of the Wellington Teachers' Training College based at Hutt Valley High School. This was part of her final year of studies for her degree in education and psychology. I could now cycle to work at Shed 2, and Rosemary took the bus to the training college. We did not have a car or the money to buy one.

I was working long hours on Claude's programme as well as with Arnold Bainbridge, but at least I did not have to contend with the long commute to Wellington. While Rosemary was studying for her final university exams and beginning her teacher training, I had extra time at night to read the scientific journal articles and reports recommended by Arnold. I still had to endure time in Claude's tritium lab, but I was learning more about atmospheric CO_2 every day, discussing its significance with Arnold and Athol after my daily work in the tritium lab was completed.

As I scaled back my involvement with Claude, I began to work more and more with Skip and Bob, the American technicians on the Scripps CO_2 programme. Initially this involved building and testing automation and data-logging equipment designed by Arnold; in 1970 this equipment was not commercially available so we had to design and build everything from scratch. I'd developed an early interest in electronics through Dad, and this – along with the problem-solving techniques I had honed during my university research projects – turned out to be vital in the development of the Makara equipment. Many of the problems we encountered were unexpected and required going back to the basic physics underlying the measurements. It was exhilarating, exhausting and absolutely absorbing. I threw myself into the programme.

The site Arnold chose for the first measurements was the Makara Radio Station on the west coast of the North Island, overlooking the Cook Strait. It had been used by Athol for his atmospheric carbon-14 measurements for more than a decade, and seemed to be an ideal position to locate the Scripps CO_2 monitoring equipment. The station was a large concrete building surrounded by aerials and equipment used to tune in to the BBC World Service, which was recorded and re-broadcast by local transmitters around New Zealand a couple of times a day.

The CO_2 monitoring equipment was set up in a basement room underneath the main radio complex. Air from the top of a 10-metre mast outside was pumped into the room, passed through a freezer plant to take out excess water vapour, and channelled into an infrared analyser, where its CO_2 concentration was measured. The analyser was a modified unit originally designed to measure dangerous levels of gases in mines. To figure out the exact concentration of CO_2 in the air, the analyser needed to be calibrated. This was done using a suite of gases that had been carefully calibrated in the Scripps lab in California and shipped to us in high-pressure steel gas cylinders. These were connected to the analyser using an automation system, so the calibration gases could be compared with the air at Makara. The concentration of CO_2 in the air was calculated by comparing it with the response of the analyser to the calibration gases.

The work was technically and scientifically complex and required huge attention to detail. I had to learn to programme some of the first computers in New Zealand to analyse the data. One of these was a PDP-8, a mini-computer first released in 1965 with a memory made of magnetic rings held in a wire lattice. The automation we designed to run the CO_2 analysis equipment at Makara was based on TTL integrated circuits – the forerunner of today's microprocessor chips. This required a knowledge of computer logic at a fundamental level, literally an understanding of how a computer electronically stores information as 'ones' and 'zeros' in a circuit made of two transistors called a JK flip-flop. In today's world of high-level computer languages, such detail would be unthinkable; most computer scientists today would have no idea how the actual individual circuits driving automation work. Yet in those days, when modern data-logging equipment was unavailable, this is what I had to get to grips with. It was an incredible time for me and I was on a high.

But it was difficult to discuss with others what I was doing. When I talked to people about my climate change concerns, I usually encountered polite indifference and even amusement. Rosemary and I went to a lot of parties in Wellington and when people found out what my job was, they shook their heads in confusion. I would tell them I measured CO_2 in the atmosphere produced from burning fossil fuels, but I had to admit that there was only a minuscule amount there and that, despite the CO_2 measurements by Dave Keeling in Hawai'i linking this to atmospheric heating, there was no actual sign that the global temperature was increasing. If anything, global temperatures had dropped slightly between 1940 and 1970, and some geologists even warned about the possibility of another ice age. When I talked to people about political disinterest in the potential danger to climate, they simply laughed. I remember one of my friends telling me about the money that needed to be spent fixing up roads first 'before worrying about stuff like that'.

But the more I studied the scientific literature and talked to Arnold and Athol, the more concerned I became. Over beer and chips at Arnold's house, he and I would talk into the night about the oceans and the changes to its carbonate system with consequences for corals and shells. He never seemed to tire of my questions. The physical science linking global warming to increasing atmospheric CO_2 was unequivocal, and there should have been a signal visible in the global temperature records. But there wasn't. Arnold had no idea why – it should have been there. It took more than a decade before, in the mid to late 1980s, global temperature readings increased and we discovered that warming had been masked by an effect which came to be known as 'global dimming'. Local air pollution from industry and agriculture had led to a shielding effect that helped to offset global temperature rises from the increasing atmospheric CO_2.

During the few months that I worked with Arnold in 1970 and early 1971, it seemed to me that he was in control of the entire Scripps atmospheric CO_2 project, something I never thought odd at the time. Later it transpired that Dave Keeling had been on sabbatical and, during his absence, Arnold was trying to wrest control of the project. It turned out that Arnold had been approaching the US National Science Foundation and other agencies for support in what became an ugly dispute between two highly talented scientists.

After I'd been working with Arnold for a couple of months, he began flying back to California, disappearing for weeks at a time. This was a disappointment. I had been reliant on him for background knowledge and context to what we were doing with the atmospheric CO_2 work, as well as for a lot of the equipment development. He was one of the most capable scientists I've ever worked with, and I felt a huge loss when I found out that he'd decided to take another job at Scripps with the GEOSECS oceanographic programme, leaving the atmospheric CO_2 project altogether. His absences meant I had to work more closely with Skip and Bob. Technically they were excellent and we made good progress with the analytical equipment at Makara, or as Skip called it, 'Macaroon'. Despite their hands-on expertise, it soon became clear neither of them had much idea about the significance of the science or the importance of the New Zealand measurements. My questions, like 'What is the expected rate of atmospheric CO_2 increase?' or 'Where is the excess CO_2 going?', went unanswered. Bob was a keen fisherman and Skip an experienced hunter. On their American salaries, life in New Zealand was pretty good. Skip was often away for days on hunting expeditions in remote places and Bob would go off on fishing trips around the country in his brand-new four-wheel drive.

We socialised quite a bit after work at a local pub in Lower Hutt, and had barbecues at a very nice apartment that Skip was renting with his wife, and at a big house Bob had taken a lease on. Both of them would talk about what to me seemed an astonishing way of life in California. They had grand ideas about 'improving' New Zealand, which included bringing in laundromats, twenty-four-hour shopping and malls. I was still very much the Taranaki kid and was impressed by a lot of their rhetoric, but Rosemary – who had been to high school in the US on exchange – discounted a lot of what they had to say. At that stage Rosemary and I were living in a tiny, decrepit two-room flat and there was no way that we could host Skip and Bob in return – they were incredibly wealthy compared to us and used to laugh about aspects of our life they found strange.

Skip could imitate my accent almost perfectly. He had a lot of hilarious New Zealand stories, including a visit to a dairy in Lower Hutt where he'd asked for sour cream and the insulted store owner replied, 'Our cream is fresh!' The concept of putting sour cream on baked potatoes was still a decade or two away for us Kiwis. The New Zealand of 1970 had a limited selection of restaurants, most of which served bland food. Fast food was confined to fish and chips and pies; we had yet to learn of KFC, McDonald's or Pizza Hut.

With Arnold away, and Skip and Bob gone a lot of the time, I was increasingly left alone at Makara, spending days and often nights just trying to keep everything running. The nights were the worst:

to save time I would often run back-to-back calibration series on the analytical equipment overnight, each run taking several hours to complete. While I was waiting for the sequences to finish, I would try to get some sleep on a mattress I had laid out on the concrete floor in a corner of the equipment room. On one occasion there was a storm overnight, and a tree blocked the Makara Radio Station access road, trapping me there for the rest of the night. There were no phones, so although I would warn Rosemary that I would be working a long day, I had no way of telling her exactly when I might get home. She was completely absorbed with her university studies, however, and seemed unconcerned by the hours I was working. I didn't see the signs then, but it was clear we were already beginning to live separate lives.

On one occasion when Arnold came back from a trip to San Diego, he found Skip in the Shed 2 atmospheric trace gas lab filing a large piece of metal in a vice.

'What's that you're making?' asked Arnold.

'A barbecue pot,' replied Skip.

'What about the wire wrap for the automation?' shouted Arnold. 'You should be working on the electronics for Makara!'

Shortly after this, Arnold sent Skip back to Scripps to work on the GEOSECS oceanographic programme. Neither of them ever worked on Dave Keeling's atmospheric CO_2 programme again. Only Bob was left, but he became more and more distracted with fishing and a student project he had been working on with Arnold through Scripps. At the beginning of 1971, to provide data for his project, he went to Antarctica on a naval research ship for a couple of months, leaving me completely on my own and responsible for the entire New Zealand atmospheric CO_2 programme.

I was lonely, spending hours on my own with no one to talk to about what I was discovering, what problems I faced, let alone what the priorities for the research were. My letters to Arnold at Scripps went unanswered. I knew that I had the equipment running well and was making the first ever continuous atmospheric CO_2 measurements in the mid-latitudes of the southern hemisphere. But the data did not make sense. During the day the CO_2 concentration would often drop to about 315 ppm then increase at night to well over 320 ppm just before sunrise. I wondered whether this daily pattern was caused by changes in the absorption of CO_2 with different temperatures in the sea surface offshore. I did some theoretical calculations that showed this was impossible. No one at the INS had any idea what could be causing the effect, and I had no one to turn to. I was convinced that something was very wrong with the data, and felt a failure. I'd read the limited amount of relevant scientific literature available and could find no clue as to what the issue was. Back then of course there was no internet to search for answers, and scientific journals arrived by sea months after they had been published in the northern hemisphere. As the only person in the southern hemisphere making continuous atmospheric CO_2 measurements, it was a worrying and lonely time.

In March 1971, I had a phone call from Athol Rafter.

'Dave, there's an American from Scripps coming to look at the CO_2 programme out at Makara. He works for Dave Keeling.'

Finally, someone who might have some answers. The American's name was Peter Guenther, and Athol told me to pick him up from the airport. But I had no idea what he looked like!

'No problem,' said Rosemary. 'Almost all young American men have crew-cuts.'

The next day I waited as the passengers came off the flight. None of them had crew-cuts. After all the other passengers had gone, I was left standing by the baggage claim beside a tall young man with shoulder-length hair, square-cut John Lennon glasses and carrying a cloth bag embroidered with coloured beads. In his shirt pocket I could see a pipe and a packet of tobacco. This was Peter Guenther. Straight away he was asking me questions about New Zealand and our way of life, and seemed genuinely interested in everything I told him – I instantly warmed to him. We drove out to the Hutt Valley in my old Austin A40, which he told me was the smallest car he'd ever been in. Communication was difficult – I had to repeat myself often because it was the first time he'd heard a strong New Zealand accent. I discovered that Peter was on a kind of spy mission for Dave Keeling. He wanted to find out what was going on with the Makara data and what Arnold Bainbridge had been up to. And also, what was this Dave Lowe character like? Peter explained the major difference of opinion between Arnold and Dave Keeling, which also explained why Arnold hadn't replied to my letters. This was the first time that I had learned anything in depth about Keeling. Up until this point, I had not even known that it was Keeling's programme that I was running at Makara. Arnold had told me nothing about him, nor about Peter Guenther. Just thirty minutes with Peter explained things that had puzzled me all year. I had literally been in a communications black-out.

I found Peter a really cool dude. He had a master's degree in chemistry and really understood atmospheric CO_2, calibration and how the measurements were made. He'd trained under Dave Keeling for two years and was fully aware of the potential significance of the Makara data.

'I assume you'd like me to book you into a hotel?' I asked Peter when we arrived in the Hutt Valley, not far from the INS.

'Uh, no – if it's okay with you, I'd rather stay at your place,' he said.

Rosemary and I were still living in the little two-room shack that was around a hundred years old, one of the original shepherd's huts in the Hutt Valley. There was peeling paint and mould on the walls, and the plumbing and hot water system were highly erratic and ancient. How was I going to put up a high-profile American scientist in such a place? Later I discovered that Peter lived in a rural commune and was not fazed at all. In fact, he welcomed the chance to live with New Zealanders rather than stay in a hotel.

'Well, if you don't mind sleeping on the couch, that's okay with me,' I said.

'That'll be great. I'm used to camping and sleeping rough.'

When we arrived, the first thing Peter looked at was my record collection.

'You like Eric Clapton?' He held up an early Cream album.

'Hell yes!'

We hit it off right away. Here was someone who enjoyed the Who, the Rolling Stones and most of the bands I liked – he'd even seen some of them live in concert. Where Bob and Skip had told me how primitive things seemed in New Zealand, Peter was interested in everything, especially the environment. He'd already read a lot about New Zealand birds and knew about the walking tracks around the country. 'This is my kind of American!' I thought.

I'd managed to track down Bob, who was on his way back from a trout-fishing trip about 400 kilometres north of Wellington. The next day I took Peter to the Shed 2 lab where he and Bob sat talking and smoking for hours. For a non-smoker this was really unpleasant, and I wondered what impact the smoke would have on the sensitive equipment. Smoke in the air, mercury on the floor and asbestos and rats in the roof! Finally, I complained.

Peter put out his pipe. 'You're right. Any chance of a cup of tea or coffee?'

I led Peter to the staff cafeteria, leaving Bob shrouded in smoke. The cafeteria was a large dingy room with grime-encrusted walls and flies buzzing round broken lightshades that were probably originally installed by the American army in World War Two. The other people there were immediately fascinated by Peter. Until then my experience with Americans was with Arnold, Bob and Skip, who would not set foot in the cafeteria.

Peter spent three days with us in the Shed 2 lab and out at the Makara sampling site. He asked a lot of in-depth questions about the CO_2 programme and my thoughts and feelings about the science. I was elated – here was someone who was really interested in what I was doing. Peter had no answer for the strange diurnal variation I was measuring at Makara but promised he would get back to me.

*

A week after Peter left, I received a phone call that I remember vividly, almost word for word to this day. It was 8am, and I was in our tiny shepherd's hut just about to head out to Shed 2 and Makara. I was not believing what I was hearing.

'Hello, David Lowe?' enquired a pleasant American accent. 'This is Charles David Keeling from Scripps Institution of Oceanography, California. I understand from Peter Guenther that you've been running my atmospheric CO_2 programme in New Zealand?'

I was speechless.

'And he tells me that you have seen a lot of variability and a diurnal signal in the data,' he continued. 'Peter is very impressed – he is convinced that the equipment is running perfectly. Thank you for that. My opinion is that the signals you're seeing have to do with the site. You must be measuring local as well as offshore effects. How far is the air-sampling mast from the edge of the cliffs above the sea?'

'Almost exactly a thousand metres,' I answered. I'd often walked from the radio station out to the cliff edge while waiting for the equipment to complete calibration sequences.

'And what is the vegetation like between the sample mast and the sea?'

'It's just standard New Zealand farm pasture,' I said. 'A mix of rye grass and clover.'

'That's almost certainly not good,' he said. 'But I would really like you to prove that the pasture is the problem and also think about finding a better site. I'll send you details on some experiments that I'd like you to run. Are you okay with that?'

I couldn't believe I was getting scientific direction from the world authority on atmospheric CO_2. Of course, I was okay with it.

'By the way, I can't afford to keep Bob Williams in New Zealand,' he said. 'I am going to have him return to Scripps. I understand from Peter that he's been living like a baron in New Zealand and often leaving you on your own with my equipment. I know this must have been very hard. I'll be in touch. Bye for now.'

I put down the receiver, my first ever international phone call, and thought about what Dave Keeling had told me. A lot of things were falling into place. Peter had clearly made a candid assessment of the situation in New Zealand, my role in running the CO_2 programme, and Bob's lifestyle. Science money is hard to get and the extravagance had annoyed Dave Keeling. And I had been vindicated in my assessment of the data from Makara: although the equipment was working perfectly, the results were compromised due to local effects masking baseline levels of CO_2.

After this, things moved rapidly. Bob was pulled out of New Zealand, leaving me on my own at Makara. Communication improved. Dave Keeling insisted on weekly progress letters from me, and I received weekly replies mostly from Peter Guenther or sometimes other Scripps staff. Dave Keeling phoned and wrote to me on many occasions over the following months, with explanations and requests. I jumped at the chance to learn from him. The work included running a test with a portable CO_2 analyser at an old World War Two gun emplacement on the cliff edge at Makara. The test took several days to complete but showed conclusively that the diurnal problem I had seen was caused by photosynthesis in the pasture drawing CO_2 out of the atmosphere between the air-sampling mast and the sea. It seems so obvious now, but at the time no one had been able to tell me that local photosynthesis in grass would cause such a huge effect on the data. Oddly enough, despite his brilliance, even Arnold had not expected this. Climate science is incredibly complex, involving many disparate disciplines – no one scientist can be on top of all areas involved in the impacts of CO_2 and other greenhouse gases on the Earth System. After I had shown that the grass pasture was causing the diurnal effects, I was even contacted by biologists who wanted to use the Makara CO_2 data to measure the onset of photosynthesis at different grass temperatures in different seasons.

My workload was immense and I continued to spend nights out at Makara. This wasn't great for my marriage. Rosemary and I were both tired and often short with each other. Our close relationship began to suffer. I was so single-minded about the Makara project and so consumed by my own work that I had no appreciation for Rosemary's. I began to make disparaging remarks to her and her friends about teaching and what a waste of time I considered it. I was convinced there was more that could be done with a university degree than teaching, stupidly not reflecting on how much I personally owed to my own mentors. But work had taken the focus away from my marriage and my idiotic remarks about teaching made things worse.

As if I didn't have enough to do, I decided to enrol for a master's by thesis at Victoria University to learn more about the factors limiting absorption of CO_2 in the oceans. I knew that one of Dave Keeling's main reasons for measuring atmospheric CO_2 in the New Zealand region was to discover the role of the southern oceans in removing extra fossil fuel CO_2. He'd surmised that about half of CO_2 emissions were being taken up by a massive sink like the southern oceans. The university and Athol Rafter were very supportive, and Dave Keeling was interested but his advice was limited to phone calls and letters. The research was basically unsupervised and I had to develop the entire project myself. It involved determining the efficiency of the absorption of atmospheric CO_2 by seawater. The amount of extra CO_2 remaining in the atmosphere is critically dependent on the efficiency of this circulation system, but at that time little was known about the process. I ran a series of basic experiments to find out how quickly a series of coastal seawater samples could absorb atmospheric CO_2, and came up with a value consistent with theoretical calculations. My thesis was submitted and passed in early 1972. It was another milestone, but I wondered whether my examiners had any idea of the potential value of my work. I suppose none of us knew then just how significant the study of ocean acidification would become. In the twenty-first century, we now know that the combined threats of increasing ocean acidification and ocean temperature increases are leading to massive coral bleaching events: by 2020 it was estimated we'd lost 50% of coral reefs over the past twenty years, with more than 90% expected to die by 2050.

The introduction to my master's thesis, which I would have written in 1971, shows that I was already very worried about what we now think of as climate change:

> Since the appearance of man on Earth the atmosphere and the oceans have had to bear the burden of his waste. Until the 1940s these sinks appeared to have an infinite capacity for assimilating waste products [. . .] but less attention had been paid to increasing CO_2 in the atmosphere due to the wide spread combustion of fossil fuels [. . .] even a relatively small temperature change could significantly alter the earth's present climatic conditions probably causing substantial melting of the polar ices caps with a subsequent rise in sea level.

By the beginning of 1972, I could begin to evaluate the CO_2 data from Makara. Due to the diurnal effect of the grass, the dataset was very 'noisy', but I could see that the minimum atmospheric CO_2 was about 321 ppm at the end of 1970 and the maximum about 322 ppm in October 1971. Sadly, Earth's atmosphere has never seen such low values again. My research showed the inexorable rise of atmospheric CO_2 in a remote part of the southwest Pacific, duplicating what Dave Keeling was seeing in the northern hemisphere. I remember being really upset by this and talking the situation through with colleagues at the INS. Athol in particular sympathised with my feelings and encouraged me to keep going with the measurements. He knew how important they were.

By March 1972, it was clear that I needed to find a new sampling location that was not influenced by local vegetation. With this in mind, Dave Keeling asked me to look for another site and send him topographic maps from around New Zealand. First, I used maps and wind trajectory data to search the Wellington area. Although I was receiving no scientific advice on running the programme, going out of the region would have made it difficult to access the basic electronic and mechanical workshop at the INS that I would need to support my work. I'd ruled out several potential sites in the Wellington region, but eventually, on a tip-off from an old friend, I decided to check out the Baring Head lighthouse, situated on a southeastern point of New Zealand's North Island at the brink of a steep cliff overlooking the Southern Ocean. Wind strength and direction data which I had obtained from the New Zealand Meteorological Service indicated that the site was exposed to air representative of a large part of the southwest Pacific. But southerly winds occurred only about 35% of the time. The rest of the time the winds were from the north, over pasture land, and would almost certainly show the same effects of photosynthesis that I had seen at Makara.

I was a bit put off by this but, when I sent Dave Keeling the wind data and maps, he was very excited. He pointed out that his site at Mauna Loa also had problems with contamination from the nearby volcano, as well as during upslope winds. However, he was able to separate out these incidents and select baseline data only. He was convinced that the same approach could be used for Baring Head if we could get permission to sample there.

The phone number for the Baring Head lighthouse was in the Wellington phonebook. The lighthouse keeper, Alan Martin, was very helpful and invited me to visit Baring Head the next day, telling me that the key to the locked gate on the lighthouse property was under the second rock to the left of the entrance gate. After crossing a river on a rickety old wooden bridge, the access road to Baring Head lighthouse was a rough gravel track winding steeply uphill, with a hair-raising drop-off on one side. The remainder of the track passed through windswept pasture land and a belt of trees marking the perimeter of the station. Without exception the trees were bent over, twisted into grotesque shapes by the constant winds. Because of the Cook Strait, a wind tunnel effect increases wind speeds at Baring Head and makes it one of the most exposed places in New Zealand; in all the years I've visited, I've never experienced a calm day there. As I drove up the track that day, gale-force salt-laden winds pounded the trees and the buildings, and I noticed the lighthouse keeper's garage was braced by huge struts and cables to prevent it from blowing away. Alan met me on the site, offered me a cup of tea and spent the morning showing me the buildings, including an ex-World War Two observation post right on the edge of a cliff. The building had mains electricity and a 15-metre flagpole. Assuming that we could measure CO_2 only during southerly winds, the site looked ideal. During my visit Alan phoned the lighthouse service for permission, and they were only too happy to see the old building being used.

The next day, I called Dave Keeling and told him the good news.

'Well done,' he said. 'It sounds like a swell place. But I have a question for you. How would you feel about coming to Scripps this year and working with my team for six months?'

He said he could send me an airline ticket to join him in California and prepare the equipment for Baring Head. People from his group would help install everything later in the year.

'This will give us a chance to meet,' he said. 'I'm very keen to talk to you about my programme and your master's thesis. What do you think?'

CHAPTER 4

SCRIPPS INSTITUTION OF OCEANOGRAPHY, CALIFORNIA

When I told Rosemary about the invitation to work for Keeling at Scripps in California for six months, she had mixed feelings.

'I'm sorry but I can't go with you,' she said. She was finishing her secondary teacher's training and, to complete her qualification, had to teach at a New Zealand high school for at least six months.

'But Keeling has promised to pay your fare to California as well.'

'No way,' she said. 'I've only just started teaching at Wainuiomata High School.'

She was right, but I was disappointed and continued to make disparaging comments about teaching. We were each so committed to our different careers that they were becoming more important to us than our marriage.

Keeling sent a checklist of things to do before leaving for Scripps. This included shutting down the project at Makara, storing the equipment in a basement room with a small heater to keep it dry. He also wanted me to bring detailed information about Baring Head, especially topographic maps of the area, wind trajectories from the Meteorological Service and photographs of the site. And he was keen to see a copy of my master's thesis too. All of these preparations, including getting a passport and working visa for the US, took months.

My departure date was set for 1 June 1972. My Pan American airline ticket was an envelope full of brightly coloured coupons representing individual legs of my flight from Wellington to San Diego. In the early 1970s, flying to the US from New Zealand involved 'island hopping' in old DC8 aircraft: Auckland to Fiji,

Fiji to American Samoa, Samoa to Hawai'i and then on to the US mainland. Climbing the steps of the plane in Auckland I was amazed when a telegram from Mum and Dad was handed to me by the cabin crew. They were very proud: Dad kept saying I was going 'stateside'. I was wildly excited – my first trip overseas. Travel to the US was rare then, and Athol Rafter was the only New Zealander I knew who had been there.

Keeling had decided I should visit his Mauna Loa measuring station on the way to San Diego, so I flew from Honolulu to Hilo, a town on the 'big island' of Hawai'i. The island is dominated by the two towering volcanic cones of Mauna Loa and Mauna Kea, each well over 4000 metres high and sacred to Hawai'ians, playing a large part in their folklore. I was shattered after a forty-hour island hopping journey across the Pacific. But my initial view of the black lava-covered cones of the volcanoes was striking. What a place – stark beauty well above the atmospheric boundary layer, a perfect site for baseline CO_2 measurements.

In Hilo I was met by John Chin, the technician responsible for the equipment on Mauna Loa. He took me up to the observatory and I was astonished at the change in the landscape. Lush tropical vegetation at sea level changed to bleak highlands above 1500 metres and black lava fields with no vegetation at all above 2500 metres. Keeling's equipment was in a US National Weather Service building at 3400 metres, and I was fascinated to see the same analytical gear I had been struggling with at Makara: a non-dispersive infrared gas analyser, automation systems, calibration gases in steel cylinders, and water vapour freeze-out traps; all were very familiar to me. The output from the gas analyser measuring atmospheric CO_2 was displayed on a chart recorder.

The record of ever-increasing atmospheric CO_2 at Mauna Loa has come to be known as 'the Keeling Curve', and is without doubt the most important long-term continuous geophysical measurement ever made. The first measurements began in 1958 and have continued to the present day, the rate of increase faster and faster as humanity's insatiable appetite for fossil fuels has grown. The simple, jagged, ever-upward trending curves from both Mauna Loa and Baring Head and now many other sites worldwide require no deep scientific training to understand: they carry a chilling warning that carbon emissions are modifying the entire atmosphere of the earth.[7]

I visited the Mauna Loa observatory again thirty-five years later in 2008, on the fifty-year anniversary of Keeling beginning his measurements there. The atmospheric CO_2 growth rate had then doubled to around 2 ppm per year. Now, in 2021, the average growth rate is well over that. The Keeling Curve carries a warning: at the very time we should be reducing carbon emissions to avoid the worst effects of a climate emergency, they are increasing at an exponential rate. What will it take for we as a species to take heed of a warning first made more than half a century ago?

After a couple of days in Hawai'i, I flew on to San Diego Airport, where I was met by Skip and his wife Pam, as well as Peter Guenther. Skip and Pam lived in Encinitas, about 30 kilometres north of Scripps, and had offered to put me up at their house for the six months I was to stay in California.

From the beginning I felt overwhelmed. The sheer size of cars, houses, freeways and buildings, not to mention the obvious wealth, was staggering. In Los Angeles Airport I'd even seen a machine that sold cans of beer for a quarter (25 cents), and I just had to try some. Unknown to me the machine was not selling beer as I knew it but root beer, a kind of American soft drink that is definitely an acquired taste. I almost threw up!

Encinitas was a small coastal village just west of a new freeway called Interstate 5, linking San Diego and Mexico in the south to Seattle and Canada in the north. Skip and Pam's house was away from the freeway traffic noise and only about a kilometre from steep cliffs with walkways leading down to a beach. The house had low ceilings and was built in an Adobe style, designed to keep out the summer heat. Everything felt strange but Skip and Pam and their other guest, Sheri Medler, made me welcome.

North County San Diego has a wonderful climate: in the whole six months I was there, it never rained. It's essentially a coastal desert with several million people completely dependent on water piped from the Colorado River hundreds of kilometres away. Inland, the landscape turns into hills covered with giant granite boulders, dry brush called chaparral, and pungent semi-desert woody plants including manzanita and various kinds of sage. Beyond is the Anza-Borrego Desert, a stark landscape that I never grew tired of visiting with Peter Guenther, although summer temperatures were often well over 40°C. The coast was much cooler thanks to coastal breezes. However, during offshore winds – Santa Ana conditions bringing desert air to the coast – temperatures could rapidly rise. Skip and Pam's place had no air conditioning, but the thick Adobe walls kept it relatively cool inside, provided the doors and windows were closed.

During my first weekend in California, Keeling invited me to dinner with his family at his home in Del Mar, a small coastal town about halfway between Encinitas and Scripps. I was incredibly nervous to meet him. Even at this early stage of discoveries in atmospheric science, Keeling had generated enormous respect in the scientific world and was well known internationally for his work. But immediately he made me feel at home, regaling me with his hilarious impressions of New Zealand from a visit he'd made in the early 1960s. He'd found New Zealanders friendly and helpful, but the weekend closure of shops and restaurants inconvenient. He also told me about the hikes he had done in the South Island along the shores of Lake Manapouri to Doubtful Sound in Fiordland, a paradise of waterfalls, lakes, fjords and dense native bush. These had clearly made a lasting impression.

Another conversation I recall from that night was of the population increase around Del Mar, which had led to increased road traffic and the building of new freeways. Keeling exclaimed that some Californians are hell-bent on what to them seems an inalienable right to drive cars. 'Some families own three or four cars and live in huge houses with triple garages!' An appalling amount of smog was generated by the millions of car engines. In the early 1970s catalytic converters did not exist, and under the sunny skies of cities like Los Angeles and San Diego, the photochemistry of exhaust emissions created a noxious brew of smog. A brown haze drifting down the coast from Los Angeles was becoming more visible each year. Keeling hated it, and I mentioned how noticeable it was to me, who was used to the clear air of Wellington. It had been a shock the week before, when my flight descended into the thick soup of smog surrounding Los Angeles. I'd immediately noticed the smell and an eye irritation when I walked between terminals at the airport. During my six months in Southern California the smog was almost always present, especially looking north towards Los Angeles. I was told that on clear days it was possible to see Santa Catalina Island off the California coast, but I never saw it.

I left Keeling's house that evening with mixed emotions. I was excited, yes – here was a chance to make a contribution to science by working with the world's foremost expert on atmospheric CO_2. But I was also affected by his overall sadness and sense of foreboding. He had an extraordinary appreciation for the environment and understood my own feeling of connection to the atmosphere and oceans. I sensed his yearning for a simpler life and what might have been – on his visit to New Zealand in the 1960s, he'd wondered about emigrating.

In California I saw, for the first time, the effects of a large and rapidly growing population on the land and air. I saw huge cars – 'yank tanks', we used to call them in New Zealand – and freeways, massive houses and buildings, everything shiny and new. But it was clear to me that an unchecked large-scale thirst for wealth and for 'things' was not going to end well. In New Zealand we lived with fewer possessions, and getting by day to day seemed more difficult. The obvious wealth in California led to what appeared to be an easy life. Was it? Many Californians had large debts on multiple credit cards paying for this ease. In 1972 credit cards did not exist in New Zealand, and what we owned was what we owned, or perhaps borrowed from friends or parents. Despite my very different background, I could imagine that if I lived permanently in Southern California it would be difficult not to succumb to the temptations of credit-fuelled buying frenzies, a condition joked about as 'Affluenza'.

Keeling knew what was coming: unchecked growth in the production of cars and material things; a consumer economy pumped by advertising, growth derived at the expense of the earth's finite resources; an economic model dependent on growth, inextricably linked to increasing CO_2 emissions from soaring fossil fuel combustion. By this stage Keeling had been warning about increasing atmospheric CO_2 for at least ten years.

Over the next six months at Scripps, Keeling would often invite me into his second-floor office cluttered with overstuffed filing cabinets and bookcases in every corner. I was in awe – this was a place where I could learn so much from this exceptional scientist. In his office, among other issues, we would discuss preparation of the Baring Head equipment and techniques needed to make the southern hemisphere baseline measurements. He was clearly excited about the site and pointed out that during southerlies at Baring Head the air would probably come from at least 55°S and be representative of a large part of the southern hemisphere mid-latitudes. He congratulated me for persevering with the Makara station and pointed out that I had already obtained a useful initial value for the baseline atmospheric CO_2 concentration.

He'd also read my master's thesis carefully. He often asked me detailed questions about my studies and the university projects I'd worked on. I was elated; all the trials and heartbreak at Makara were shown in a positive light. I set my mind on learning as much as possible from him and doing the best I could during my time at Scripps. He was also interested in everything I could tell him about New Zealand, particularly the politics, economy, education system and science coming out of Athol Rafter's laboratory in Lower Hutt. Keeling was a brilliant man with a reputation of being hard on his staff, but I found him inspirational. Although I was a young scientist, he treated me with respect. His mentorship and encouragement changed my life forever.

Keeling approved of my love of surfing provided it did not compromise my work. I used to store a surfboard in one of the labs at Scripps and, whenever the surf was running and my workload allowed, it was only a 100-metre dash down to the water's edge. I felt on a high: I was working with the top experts in my field and could go surfing whenever time permitted. The months flashed past. I was so involved with my work and life in California that I rarely thought about New Zealand, my parents, my friends, or even Rosemary. Phone calls to New Zealand were prohibitively expensive, so contact was limited to occasional aerogrammes and postcards. I was having the time of my life.

'Surf's up, Dave!' my colleague Carl Ekdahl would yell across the lab when he wanted to hit the waves. The only other surfer on staff, Carl was an experienced physicist who worked on modelling studies of atmospheric CO_2 as well as analytical equipment. I got on with Carl right away and enjoyed talking to him about the latest advances in physical science, especially solid-state physics. He was proud of his distinctive waxed moustache which stuck out at right angles from his face. The rest of the staff called him the Red Baron, after the ace German World War One fighter pilot. He was a keen glider pilot, and once took me on a flight over the desert. We worked hard but enjoyed taking time off if the surf was running and duties in the lab allowed.

My surfboard slapped the waves as I paddled out through the breaks north of Scripps pier with Carl one day. All around me were the excited voices of other surfers as we waited for the next set of waves. It was strange at first to be surrounded by American accents. Now, after three months, I was used to it and the perennial drawled question, 'Where y'all from?' when anyone heard my accent. Back then most Americans had never heard of New Zealand let alone knew where it was – in fact I'd seen several world maps that did not even have New Zealand on them. Often, I would describe New Zealand as a small country in the southern hemisphere. But the concept of Earth being split into two hemispheres still confused some. Usually I settled for 'New Zealand is a small country about halfway between Vietnam and the South Pole', deciding that most Americans would know of Vietnam because of the war. I remember one person asking me if New Zealand was a sort of mid-Atlantic island where aircraft refuelled on flights between the US and Europe!

Carl told me I should try surfing the break off Encinitas called Swami's after work sometime. Skip had also told me about this spot, a reef break not far from his house across a set of railway tracks and a four-lane highway. Because of the difficulty getting a surfboard to the spot, I decided to leave the board behind and see whether I could body-surf the break. When I got to the beach, I could see waves breaking on the reef about 300 metres out. I swam out and found myself completely alone in water at least 5 to 6 metres deep beyond the reef, well out from the beach. I'd caught a few good waves and was starting to tire when I saw someone swimming towards me from the beach. I was astonished when a bearded man with long spiky hair and a strange look on his face swam straight up to me.

'Whom do they say that we are?' he shouted, pushing one of my shoulders down.

I was too surprised to answer.

'Whom do they say that we are?' he yelled even louder, pushing me again.

This happened another couple of times. If he became violent, I was going to be in serious trouble. A number of possible scenarios flashed through my mind, none of which ended well for me. Feeling increasingly desperate, I suddenly remembered the effect my strong Kiwi accent had on other surfers I'd met out in the waves near Scripps.

'I'm buggered if I know!' I yelled, looking straight at him.

The guy paused, shook his spiky locks in amazement, and swam back to the beach, shouting all the way in. When I told Carl about it next day, he laughed.

'Oh yeah, sorry I forgot to warn you about that. That reef break is called Swami's because on the cliff above the beach there's this sort of Indian temple called the Self-Realisation Society. They attract a lot of weird people from all over who are trying to figure out the meaning of life and everything.'

I kept surfing at Swami's but always went to the trouble of taking a surfboard across the train tracks and the highway and down the cliff to the beach. Sitting on a surfboard by the reef felt a lot safer. Occasionally I saw people behaving strangely on the beach, but no one bothered me out by the reef again.

Living at Skip's house in Encinitas with his wife Pam and housemate Sheri was great. Skip was often away on oceanographic cruises working for Arnold Bainbridge, and I would share household duties with the two young women, paying monthly rent for my room. We soon developed a routine which had me shopping and cooking a couple of times a week for the three of us, as well as taking our clothes to a laundromat about ten minutes up the coast.

In the early 1970s there were no supermarkets in Wellington, so it was a shock when I first went into the local Safeway supermarket in Encinitas. I was completely unprepared for the overwhelming range of items, many I had never heard of. Also, some things were packed and treated differently than in New Zealand – for example, shrink-wrapped cut up oranges in a fancy box. Why, I wondered, would you not just sell the orange in its natural skin?

One of my early mistakes was buying minced turkey meat. When I made this into hamburger patties and tried to cook them one night, the patties turned into a dubious mess of melted fat. 'No one buys that stuff except dumb Kiwis,' said Pam, laughing. 'Try some ground lamb next time and check out the salads.'

The salad items were outstanding, fresh from nearby market gardens. I soon discovered avocados and Mexican food – corn tortillas with delicious fillings of refried beans, shredded meats topped with different salsas, grated cheese and sour cream. Most American beers at the time were almost as bad as the root beer I had gagged on at Los Angeles Airport, but fortunately Mexican beer was quite good and readily available in California.

We used to have a lot of impromptu parties at the house in Encinitas, replete with a truly terrible wine called Cold Duck, and Mexican beer and food. University academic staff and Scripps colleagues would attend, and I was intrigued to see them smoking marijuana. The stuff was everywhere and, though it was technically illegal, it seemed to be part of everyday life in California – Pam even used to bake it in biscuits she called brownies. The parties often turned into all-nighters, with many of the guests wiped out by a combination of marijuana and alcohol. Sheri, Pam and I would all have terrible hangovers next day and swear off the Cold Duck, but invariably we would throw another party a couple of weeks later with the same results.

A lasting memory I have of that time was the terrible impact of the Vietnam War on the psyche of young Americans. Many of the young men at our parties had been drafted, and this was a constant topic of conversation. At university in New Zealand I'd demonstrated against the Vietnam War outside parliament buildings, but that was low-key compared to what was happening in the US. I met people who were torn between loyalty for the young men doing their patriotic duty and despair at the human cost and insanity of an unwinnable conflict. Without exception the young people I met had lost friends and relatives in Vietnam, a nightmare that was to continue for another three years.

The end of each working week at Scripps was marked by an event called 'Thank God it's Friday' or TGIF, but simply 'Tee Gee' to Keeling's staff. We would make our way to the boardwalk platform above the beach at Scripps and sit drinking weak draught beer, watching the sun slide into the sea. We were a motley mix of scientists, students, surfers and potheads. Behind us a group would play an intoxicating rhythm on Congo drums joined by guitarists launching Santana riffs towards the waves. We'd look for the 'green flash', an elusive atmospheric phenomenon that occasionally occurs just after sunset, when the horizon is clear and the atmosphere happens to filter out red and blue light, leaving green as a brief flash on the disc of the sun. Only the truly stoned and drunk professed to see it, yelling 'Wow, far out man!' as if this would help the rest of us see something.

I never saw the green flash at Scripps, but there were lots of other things to wonder at as the crowd began to mellow. If there was a big surf running, we saw the waves rolling in from the horizon. They raced along the Scripps pier before breaking on the shore beneath the boardwalk. Rays of light shot iridescent colours over the Pacific Ocean and mixed with the smell of rotting seaweed and ocean spray. Behind us, a dark indigo, the foothills of the Laguna Mountains, the chaparral and the Anza-Borrego Desert. Suddenly it would be dark and clammy as the dew began to drop. The waves would light up with phosphorescence generated by millions of tiny creatures as the surf continued to pummel the shoreline. Ocean waves and the atmosphere are precious artforms which I've never grown tired of watching.

A few weeks before I was to return to New Zealand, Keeling called a special staff meeting. We perched on chairs and stools jammed in between the filing cabinets and bookcases in his office and he explained our next project. We'd been using a couple of different infrared gas analysers for our atmospheric CO_2 measurements, all of which were calibrated with reference gases containing CO_2 in nitrogen. Reference gases are an essential part of any atmospheric CO_2 analysis system because they provide a laboratory-produced, calibrated CO_2 concentration against which to compare unknown air samples using a gas analyser. At remote air-sampling sites like Baring Head and Mauna Loa, the atmospheric CO_2 concentration is calculated by measuring the difference between the air value and a series of reference gases that are stored at the site in high-pressure gas cylinders.

Keeling told us that he wanted to make comparisons between the different kinds of infrared analysers that we had been using to make atmospheric CO_2 measurements. The one I had been using at Baring Head was called a URAS-1, whereas the one at Mauna Loa was an even older instrument called an APC. In the lab at Scripps there was also a newer gas analyser called a UNOR, which at that time had not been field tested. Keeling asked me to work with Carl and oversee a parallel experiment of the three different gas analysers measuring exactly the same air sample.

Setting up the equipment to run the comparisons took a couple of hard weeks of plumbing stainless-steel gas lines and programming and installing automation, all tasks that I was used to from my experiences at Makara. Carl and I were already close from surfing together, and he trusted me to complete the preparations for the experiment. He was away when I finally began the tests. Within an hour I was astonished to see that the new instrument, the UNOR, was measuring atmospheric CO_2 concentrations above the reference gas value, whereas the other two analysers used at Mauna Loa and at Makara were producing readings below the reference gas values. This seemed impossible, because they were supposed to be measuring exactly the same air sample. How could the results be different? I immediately phoned Keeling.

'Are you sure?' he asked. We stood dumbfounded in front of the rack of equipment.

'Yes – but it's impossible,' I said.

Keeling looked visibly upset and said he had some ideas about what was happening. 'We did speculate about a possible effect caused by not having oxygen in our CO_2 reference gases. We didn't take it any further because the analysers only respond to signals in the infrared from molecules like CO_2 and water vapour, whereas simple molecules like nitrogen and oxygen don't absorb in the infrared and so aren't directly detected. But indirectly they could have an effect on the infrared absorption of CO_2. I'll need you to figure out what's going on.'

More than a quarter of a century later, in his memoir *Rewards and Penalties of Monitoring the Earth* (1998), Keeling credited me with the discovery of what came to be known as the 'carrier gas effect', requiring a paradigm shift in how we measured atmospheric CO_2. Of course, it was a team effort, particularly with Carl and our colleague Dave Moss, and I just happened to be on the spot making the first measurements.

I was impressed with the way Keeling handled what was potentially a serious problem for the integrity of all of the atmospheric CO_2 measurements he'd been making for the previous fourteen years. By its very nature experimental scientific research involves delving into unknown territory, making measurements using equipment and techniques that have often not been used before. Occasionally, cherished research turns out to be useless and has to be discarded. As well as having intellectual firepower and a rigorous approach to experiments, the best research scientists are almost always highly creative. Having to start over would have been demoralising, but Keeling immediately thought through the consequences of our discovery and devised new experiments to try.

He decided that we needed to begin using reference gases based on CO_2 in air rather than the simple CO_2 in nitrogen mixtures we had been using. This required years of repetitive calibration work by Peter Guenther and others. However, the final result was a much more robust calibration system of reference gases based on CO_2 in air that could be used by different analysers in laboratories all over the world.

By December 1972, after an incredible six months in California, it was time for me to return to New Zealand. I'd learned and experienced a huge amount, and Peter and I had completed preparations for the equipment to be set up at Baring Head. Keeling had briefed us with his requirements and advice. Once we were up and running at Baring Head, we were to stay in contact with weekly letters and keep detailed field logbooks and reports on our work.

I had mixed feelings about returning. Life in California had been wonderful. I'd grown and matured, and made friends with Americans socially and at Scripps. I realised that in New Zealand my life had been quite insular, revolving around Rosemary and my atmospheric research, with little time to think about anything else. In California I'd had the freedom to challenge myself in so many ways. After six months, it was a very different Dave Lowe returning to New Zealand – mature and wiser, and with reservations about my marriage. Rosemary and I had been apart for six months – would we feel like strangers? I also felt a sense of dread for the task ahead at Baring Head. The Makara situation had caused me so much stress – what if Baring Head turned out to be another bad site? The only way to know was to get there and start the measurements.

CHAPTER 5

BARING HEAD

Baring Head lighthouse is on a steep cliff overlooking the Southern Ocean. Between the cliff edge and Antarctica lie thousands of kilometres of open sea, the most isolated and windswept ocean on the planet. To the southwest, on clear days, the Kaikōura Ranges in the South Island are often visible, with the snowy peak of Tapuae-o-Uenuku, formerly Mount Tapuaenuku, about 150 kilometres away. Baring Head is one of the most exposed places in the Wellington region, with gales from the north as well as the south routinely pounding the lighthouse station. The few trees that survive are bent over at weird angles and dense, wiry wind-resistant shrubs and flax drape over boulders near the cliff edge. Although there is grazing land to the north, there is little vegetation between the station and the sea. It's an awe-inspiring environment, raw and beautiful.

By mid-December, Peter and I had shifted most of the sampling equipment I'd originally used at Makara to a small concrete building at the edge of the cliff at Baring Head. The 12-metre flagpole there was ideal for hoisting the air-lines we used to collect uncontaminated air for the infrared analyser we'd set up in the building.

After only a couple of days of sampling, we made the first atmospheric CO_2 measurements during a southerly wind. The levels on the chart recorder were remarkably constant, something that I'd never seen in all the time I had made measurements at Makara. After calibration adjustments we could see that the atmospheric CO_2 concentration at Baring Head was about 323 ppm, already 1 ppm up on the 322 ppm I'd estimated for Makara the year before. These were the first ever continuous measurements of uncontaminated atmospheric CO_2 from the remote Southern Ocean.

Our excitement was shortlived. When we returned to Baring Head the next day, there was an ominous silence in our equipment room. Nothing was running. The fault turned out to be a borrowed power cord which had been wired incorrectly, and had caused multiple failures in the infrared analyser, as well as the automation used to drive the calibration gas sequences. One circuit board inside the analyser was completely fried, and I had to rebuild the circuit from scratch. As Peter wrote in the logbook, 'Another hard lesson – never trust equipment from someone else that you have not tested!'

Over the next couple of months our frustrations increased. We worked at Baring Head virtually every day except Christmas Day, battling every imaginable equipment problem: failed chart recorders, automation, water freeze-out traps and air pumps. Murphy's Law – anything that can go wrong, will go wrong – was proving true.

In those early months it seemed that everything was against us. We worked incredibly hard, often spending nights sleeping on the floor of the equipment building or in the relieving lighthouse keeper's house. But it seemed that no sooner had we solved one problem than another would crop up. Electronic problems with the CO_2 gas analyser were among the worst and seemed to be a weekly occurrence. The analyser would routinely overheat, and I had to redesign the temperature controller to stabilise its operating temperature. We'd get a couple of days of useful atmospheric CO_2 data before something else would fail, like the gas analyser's sensitive infrared detector unit, and I would have to turn round and rebuild that. The violent weather at Baring Head routinely damaged and tore air-lines off the flagpole and we would have to climb up on the roof of the building and replace them, despite unpleasant conditions. The original, handwritten station logbooks bring back a vivid picture of the adversity and almost daily trials we faced. Making measurements of atmospheric CO_2 at a remote site like Baring Head is, and will always be, a challenge requiring endurance and perseverance from dedicated science staff.

Peter saw the Kiwi expression 'She'll be right' as a corollary to Murphy's Law. He reckoned if you heard a Kiwi say 'She'll be right', then everything was almost certainly not right. All these years later I'm amazed at what Peter and I put up with. It was demoralising work and often had us close to tears, especially after spending a day repairing equipment only to find another thing broken the next. But never in those terrible months do I remember Peter complaining; I've only seen him get mad once, when we realised we'd forgotten to pack beer on a trip to the Anza-Borrego Desert!

During the first few months of our trials at Baring Head, we were in constant contact with Keeling, mostly by airmail but also the occasional phone call. This is a copy of a short letter from him in March 1973:

Dear Peter and David,

Eva will be writing soon and acknowledge[s] all letters received. I had intended to write you once a week myself, but without someone to remind me, it's easy to slip. I believe I have all your letters up to and including 27 February (no. 13).

I'm very pleased with how things are going generally. It's a relief that you have been able to solve the most pressing problems. We are trying to act promptly on your requests. Let me know if you feel that we are falling behind on anything of this nature. The Baring Head data appear to be good enough to get us by for now without a second station. We are working it up now with plots, and blue line averages, since this is the month, we write the NSF proposal.

I have begun to do the finalizing on the Makara article now that the South Pole and Mauna Loa articles have been sent to Tellus. I'll send a revised copy when ready. I appreciate the promptness with which you caught up the data. You have succeeded in getting us what we need for the NSF proposal. Also, the regular weekly letters help us.

Sincerely yours,

Charles D. Keeling

Keeling continued to encourage us and acknowledge our hard work. But at the same time, he insisted on results – that we provide baseline CO_2 data to satisfy the US National Science Foundation, who were funding his programme. He needed funding to renew the programme and was having to defend his work with administrators who did not appreciate the significance of what he was doing. Keeling's autobiography reveals the frustration and petty issues this extraordinary scientist had to endure to provide the data which now form the backbone of our knowledge about increasing atmospheric CO_2 and its impact on climate change. It was incredibly frustrating for him to have his research constantly threatened by bureaucrats who had no appreciation of his work. At that stage I was still very much a junior scientist and, although I provided input into Keeling's funding proposals, I was not directly involved in their submission. Little did I know that my future career would involve similar battles with organisations often more interested in the layout and box-ticking of a funding proposal than the actual significance of the research put forward.

By mid-year Peter and I had managed to get on top of most of the equipment problems at Baring Head, and the workload was starting to ease. Doing three to four trips a week there to sort out the continual equipment problems had been exhausting. The drive was 30 kilometres each way, the last 10 on rough and dusty roads. The final 5 kilometres were little more than a farm track carved out of bedrock and mudstone, starting with a hairpin bend at the bottom of a steep hill beside the Wainuiomata River Valley. For many of the trips we used my old Austin A40, which was already in bad shape. It deteriorated further on the Baring Head trips with constant exhaust, engine and suspension problems. Holes in the chassis funnelled road dirt into the cab, and on the last section we drove with the windows open to try to avoid breathing in the thick dust. Conversation was impossible above the rattle of rocks and gravel. The trip added to the exhaustion of our long days. Peter bought an Austin A35, even older than my car, and we also used that. Somehow we managed to keep both these ancient vehicles going.

During my time in California, Rosemary had moved from the old shepherd's hut we had been living in to a three-bedroom 1930s house which she'd bought for us near the eastern hills of Lower Hutt. I invited Peter to stay with us as well as Sheri, one of my roommates from Encinitas, who was planning to meet her boyfriend in Australia. The peeling paint on the house in Lower Hutt needed urgent attention, and repainting was a condition of one of the three mortgages that we'd taken out to buy the place. The garden was full of weeds, and lots of trees at the back of the property needed trimming. Having to maintain a garden and house did not fit with my total involvement in the Baring Head project. I occasionally managed to mow the overgrown lawns and just let everything else slip.

While I was away, Rosemary had completed her teacher's training and accepted a job as a mathematics teacher in Wainuiomata, not far from the house. As a new teacher she had to put in long hours at the school and time at home marking student papers. She had little energy for me and my descriptions of progress at Baring Head. I was annoyed by her lack of interest and continued making disparaging remarks about teaching. Rosemary was becoming more and more interested in feminism – an important issue in the 1970s, with women's rights, liberation and diverse social movements gaining momentum. She became immersed in feminist literature, including Betty Friedan's *The Feminine Mystique* and Germaine Greer's *The Female Eunuch*. I tried to talk with her, but she seemed increasingly indifferent to my attempts to reconcile a worsening situation. We were on different trajectories, interested in different things and unable to communicate about what each of us thought important. I felt dread and was increasingly upset at the loss of the intense feelings we'd had for each other during our student years. After only three years of marriage, we were drifting apart. Peter and I would tip toe round Rosemary, often cooking separate meals or bringing home fish and chips after a late night.

In March 1973 Peter's girlfriend, Chris Kenyon, arrived from the US and moved in with us. Peter was still coming to terms with the tough divorce he'd gone through the previous year and did not know his mind. Chris had given up a lot to come to New Zealand and Peter was often uncaring and uninterested towards her. What a mess it was – two couples in a small house on completely different pathways and with different needs. Needless to say, the atmosphere was often charged, there were frequent rows, and I yearned for my life back in California. Like an idiot I often mentioned this to Rosemary, which of course wound her up. After a couple of difficult months, Rosemary came up with an ultimatum: Peter and Chris had a month to find another place to live. Peter was upset but soon he and Chris moved into a small flat in downtown Wellington, where they lived for the rest of their time in New Zealand.

After six months at Baring Head, with many of the nagging and persistent equipment problems fixed, we had more time to work up the data. This was a priority for Keeling. It also allowed me to finish my article evaluating the Makara measurements. Although Makara had not produced the quality of data needed for a long-term atmospheric CO_2 monitoring site, I was still able to show that the baseline level of CO_2 was 321 ppm in February 1971 and had reached 322 ppm by that October. Despite their poor quality, the Makara data confirmed the awful truth of increasing atmospheric CO_2 worldwide. This, my first article, was published in the *Journal of the Clean Air Society of Australia and New Zealand*.[8]

During the winter in 1973 there were a lot of southerly gales with an exceptional amount of baseline data collected at Baring Head. During these conditions, meteorological data showed us that the air had come from thousands of kilometres south of Baring Head, representing a large windswept area of the southern oceans completely remote from industry and vegetation. This baseline CO_2 data was the gold standard we were aiming for, and we joked that we only got it during the most atrocious southerly storms. Anemometers designed to measure wind speed were a constant problem – they often fell apart in the gales. In such conditions we would often hang over the cliff into the mind-numbing roar of the wind, watching as huge waves pounded the beach and rocks below the sampling station. The weather was exhausting and punishing, but the sheer force of it was awe-inspiring; we never grew tired of looking out over the Baring Head environment.

In addition to our measurements of atmospheric CO_2, Keeling asked us to take discrete air samples at the cliff edge during southerly gales. These samples were collected in two- and five-litre flasks (called Keeling flasks within the atmospheric community) which had been evacuated in the laboratory at Scripps before being shipped to us. The air samples were collected by hanging over the fence at the cliff edge, facing directly into a howling wind with the flask held forwards. The collection process involved holding your breath, opening a valve on the glass flask and listening to the *whoosh* as an ambient air sample was sucked into the vacuum of the flask. When the noise stopped, sampling was complete and the flask was full of uncontaminated Baring Head air. Closing the valve then sealed the flask, and the air sample was ready for analysis back at Scripps. We would collect about twenty air samples in Keeling flasks over a few weeks, and then ship them back to Scripps in specially made wooden boxes.

It was low-tech but effective. Keeling had used the technique to get some atmospheric CO_2 data from the Amundsen–Scott South Pole station, a weather observation station built during the International Geophysical Year in 1957, as well as air samples from a series of ships making oceanographic voyages. We used the flasks to check the in-situ CO_2 analyser at Baring Head. It was always reassuring to find that they, within error, provided the same values as the continuous infrared analyser we had sweated blood to install and keep going.

There were a couple of issues with collecting air samples in flasks which even Keeling had not foreseen – the first had to do with the outside air temperature during a winter southerly storm at Baring Head. The temperature would often be only 5 or 6°C and this, combined with the strong winds, would cause the greased glass valves in the flasks to lock solid. In this case, the instructions were to take our gloves off and warm the greased glass valve with our bare hands. The problem with this was our hands quickly became too cold to warm up anything, let alone a large glass valve. Peter and I improvised by using a hair dryer to warm the glass valves inside the building, then running outside to the cliff edge with the flask and taking the air sample before the grease in the valve froze again. The second issue had to do with listening for the *whoosh* as the air sample was drawn into the flask. Sometimes the wind was so noisy we could hear nothing else – we just had to guess when the flask had stopped sucking in air.

In the second half of 1973, the hard work was starting to pay off for both Keeling's Scripps CO_2 programme as well as Athol Rafter's institute. During every southerly we were collecting valuable baseline CO_2 data representative of a large part of mid-southern hemisphere latitudes. Keeling was pleased with the results, which he needed urgently to justify renewed funding for his atmospheric CO_2 programme. We were able to establish a seasonal variation in the data with a minimum in April of about 323 ppm and a maximum in October of about 325 ppm. The seasonal cycle was much smaller than at Mauna Loa and we assumed that it might be due to the lack of vegetation in the southern hemisphere. Most of the earth's landmass and terrestrial vegetation is in the northern hemisphere; the southern is very much the ocean hemisphere. It disturbed me that the Baring Head measurement in October 1973 was at least 2 ppm higher than the measurements I had made at Makara in October 1971, our measurements corroborating the growth rate that Keeling was measuring at the Mauna Loa station. Without doubt, atmospheric CO_2 was going up in the whole of the earth's atmosphere, something that both Peter and I discussed daily. I began to talk about our findings openly with other scientists at the DSIR, and for the first time these concerns began to get traction. At last, some of them were beginning to understand the significance of our measurements as the series grew and the dreadful implicit message became clear.

It was obviously important to keep the Baring Head measurements going. We knew only about half of the CO_2 emitted by burning fossil fuels remained in the atmosphere: the 'airborne fraction'. The remainder, which Keeling referred to as 'missing' CO_2, seemed to be absorbed by the planet – but where? A likely candidate was the southern oceans, and this made the Baring Head programme a priority. It was the only set of measurements at the time capable of inferring the potential importance of the southern oceans in removing excess CO_2 from the air. But to do that we needed several more years of baseline data to deduce the seasonal cycle and annual growth rate. With encouragement from Keeling, I was more than prepared to do the work – the more data I produced, the more valuable the time-series of atmospheric CO_2 at Baring Head became.

Keeling, depending on the funding he could attract, also wanted to increase the number of atmospheric CO_2 sampling stations in New Zealand. He encouraged me and Peter to look for other potential coastal sites. The New Zealand Lighthouse Service had been helpful at Baring Head, and the lighthouse keeper had been able to routinely check the equipment for us while we were away from the station. We contacted the service and got permission to visit six different lighthouses. This included Cape Reinga in the far north and Cape Egmont near New Plymouth. We also visited several lighthouses in the South Island, including Farewell Spit and Westport. At each site we made careful evaluations of the surroundings for potential sources of contamination as well as local meteorological records.

I'd thought that Baring Head might have limited value due to the relative low frequency (30%) of onshore winds from the south. But all of the other sites had various problems. We'd thought the west coast of the South Island would be perfect, but sites there suffered from a local effect caused by the Southern Alps – drainage airflow from the Alps ('katabatic winds') often brought contaminated air from inland out to the coast. This effect virtually ruled out any possible site there. After reporting our findings to Keeling and seeing the growing record of high-quality data at Baring Head, it was becoming clear that we had struck gold with our first site. Yes, we had to throw away about two thirds of the data, but the remainder provided an essential record of what was happening in the southern oceans – exactly what was needed to complement the record from Mauna Loa and the flask sample data from the South Pole station. The other advantage was that Baring Head, despite the arduous drive in our ancient cars, was handy to the well-equipped laboratories of the INS and the DSIR in Lower Hutt.

But science is not just about battling equipment and solving problems. Because you are spending public money, you are subject to the whims of people running the science funding agencies; funding can be a lottery. Usually science fund administrators are highly intelligent people and passionate about supporting top-level and valuable research. But proposals for science funding are often reviewed by other scientists, and this can cause bias when competitive programmes are being evaluated. Peter and I were junior scientists, and at that stage we did not take part in the initial writing of funding proposals for new work. Unfortunately, towards the end of 1973 both Dave Keeling and Athol Rafter struck funding issues almost simultaneously, a perfect storm. On a whim, a senior administrator at the DSIR head office in Wellington decided that atmospheric CO_2 measurements had nothing to do with the isotopic research work done at INS, and the Baring Head effort should be closed down immediately. Athol was extremely annoyed, pointing out the close linkages between stable isotopic and radiocarbon research, and atmospheric CO_2 measurements. His early radiocarbon work had shown that natural atmospheric CO_2 was being diluted by CO_2 containing no carbon-14, which could only have been produced from fossil fuel combustion. But to no avail; the administrator at the DSIR head office had no idea what he was dealing with. Thankfully, Keeling wrote a strongly worded and persuasive support letter to the DSIR detailing the international value of the programme and imploring them not to shut it down. After some discussion, the DSIR reversed its decision, especially after other organisations in New Zealand like the Meteorological Service backed the stance taken by Athol. This was my first introduction to the power that administrators remote from research hold over science programmes, making often arbitrary decisions with far-reaching consequences.

Keeling's own funding problems were potentially more serious. In the 1960s he had already lost more than a year of vital Mauna Loa atmospheric CO_2 data at the whim of an administrator in Washington, DC. But in the 1970s, the US National Science Foundation decided it would no longer fund 'repetitive research' that should be left to organisations like weather bureaus, and proceeded to seriously cut programme funding that had previously been approved. Peter and I were aghast. The work we had put in was hardly repetitive; problem after problem had arisen and our very particular backgrounds in atmospheric science had been pushed to the limit to provide the high-quality data set which was undoubtedly of great importance to the international community. Although both the New Zealand and US governments knew about the importance of the research, this did not translate into vital funding needed to continue our work, instead often run on a shoestring and supported at great personal cost to staff who put in unseemly hours to keep the measurement series running.

At this stage, Peter was being paid directly from Keeling's budget at Scripps. Keeling contacted Athol's deputy to see whether Peter could be taken on as staff of the INS to save US project funds. American salaries were substantially higher than New Zealand ones, and Keeling figured that having Peter on a New Zealand salary while working in New Zealand was entirely reasonable, even if it ignored the fact that Peter had a lot of bills to pay back in the US.

It was a stressful time. I remember us getting drunk one night, talking through the implications of Peter's new salary. He and Chris loved living in New Zealand, but what of the future? After a lot of discussion, they decided to return to California in December 1973, leaving me on my own to run a complex and time-consuming project. It was a sad and bitter time for us both. During all the adversity at

Baring Head, Peter and I had formed an unbreakable friendship, a bond that has since lasted for almost fifty years.

I was on my own at Baring Head. Problems with the equipment persisted, and at times it seemed these would never let up. I was constantly having to replace parts in the gas analyser and make modifications to keep it running. The wind and salt air continued to destroy air-lines. On top of this was the ongoing threat that Baring Head might be shut down due to funding problems. Rosemary and I spent very little time together and when we did interact, despite my best efforts, we had less and less in common. As a consequence, we barely talked to each other. Although we were in the same house, we were leading different lives – she would go out and socialise with friends while I spent what little spare time I had surfing and playing guitar. With Peter and Chris gone, I was feeling alone and full of self-doubt about my personal life as well as my research. I knew Athol Rafter, Keeling and many other scientists understood the value of the Baring Head measurements. If it was so self-evident, why was funding and the attitude of administrators becoming such a battle?

Towards the end of January, Athol called me into his office with a proposition. Scripps were running an oceanographic voyage from Wellington to Antarctica for six weeks in February and March, and had suggested I might go. The task involved running a new instrument to measure the amount of CO_2 dissolved in seawater – similar to what I'd worked on for my master's thesis. I was thrilled – but felt tied to Baring Head. Athol replied that I would train Owen Rowse to help run the programme in my absence. A brilliant science technician at INS, Owen would be ideal. There was also a helpful new lighthouse keeper at Baring Head, Barry Woolcott, who had taken a major interest in the project. He would phone me when he suspected there were issues with the equipment, something which had already saved me a lot of time and prevented the loss of valuable data. It was settled – I was bound for Antarctica.

When I talked to Rosemary about the trip, she almost seemed relieved that I would be away for six weeks. This left me depressed – I really cared for her, but there seemed to be no way I could make her happy or talk about our feelings without upsetting both of us. And who could blame her. She was young, highly intelligent and had her own ideas of what lay ahead. I was an insensitive and inconsiderate fool. I felt torn, but I was dedicated to my research work. I had to continue.

CHAPTER 6

ALONE IN A DARK PLACE

Huge waves swelled as far as I could see. It was mid-February; we'd left Wellington two weeks before and were now well south of Campbell Island, at 55°S on the way to the Ross Sea in Antarctica. The *R/V Melville* was tiny, less than half the size of New Zealand's interisland ferries, and rode the waves like a surfboard. During the frequent storms, waves would break right across the decks, occasionally reaching the height of the bridge. Skip, now working for Arnold Bainbridge and the GEOSECS programme, was on board with me. I marvelled at the professional way he supervised the deck crew, using winches to bring seawater samples from a depth of 3000 metres to the surface for radiocarbon and other analyses. Skip had been more interested in hunting than the atmospheric CO_2 programme when he worked with me at Makara two years earlier, but here he was thriving in an essential technical role. Due to frequently dangerous sea conditions, his work would be curtailed and he spent a lot of time asleep in his bunk. My work was indoors, running analytical equipment that measured gases dissolved in seawater, pumped into the lab through stainless-steel tubes attached to the hull of the ship. I shared a cabin with Skip and at times the rolling and pitching of the ship was so severe that we had to tie ourselves into our bunks with rope. Most of us were sea sick at some stage, and I was no exception.

Running the equipment was challenging but fascinating, and supplemented what I'd learned during my thesis work. Measuring the CO_2 dissolved in surface seawater was one of the first ever experiments to investigate the role of the southern oceans in absorbing excess CO_2

in the atmosphere. Fifty years later, we now know that this expanse of ocean plays a pivotal role in the earth's climate. Most of the extra heat absorbed by the planet due to its increasing greenhouse effect is stored in the oceans – current estimates are about 93%. But in addition, increased atmospheric CO_2 is being absorbed by the oceans and causing a dangerous increase in ocean acidity.

Towards the end of February we sailed into the Ross Sea, reaching 75°S. This was the southernmost part of our journey, not far from the Ross Ice Shelf and New Zealand's Antarctic station at Scott Base. It was bitterly cold, but thankfully the storms had ceased. We saw large numbers of southern right and minke whales and royal albatrosses. Penguins slid down the faces of icebergs into the sea. All around us was stark beauty.

After I returned to New Zealand, I described the experience in a letter to Mum and Dad on 8 March 1974:

It took about 10 days' sailing from Lyttelton, including three stops for seawater sampling, to reach the Ross Sea. Once there we spent about a week amongst the pack ice and icebergs. The bergs were beautiful especially the ones which had become top heavy and toppled over several times. These had colours ranging from iridescent blues to deep greens. Probably very similar to what you saw during your North Atlantic sea going days in the 1930s Dad? I also got to know the ship's radio operator fairly well and spent some time with him in the radio room. Got a feeling for what you must have done on the merchant ships in the Atlantic when you were younger. We saw a lot of wild life including one Royal Albatross which would have had a wing span of at least 10 feet.

Food on board was OK except that it was very rich, typical Yank stuff. Luckily, I could sometimes make my own when on night watches. Actually, with light 24 hours a day, night didn't have much meaning. It didn't really matter what time you went to bed – so long as you got eight hours sleep/24-hour period you were ok.

One of my most enduring memories from the trip was seeing an aurora directly overhead. Despite the twenty-four hour daylight, we could see huge shafts of light, the colours slowly shifting above us like folds of giant curtains, rippling as if moving in a breeze. Years later, a knowledgeable friend told me that, at 70°S, the southern magnetic pole would have been almost directly above us, helping to focus solar-charged particles striking atoms and molecules of gases in the upper atmosphere. At high latitudes the effect is so bright that it is easily visible during daylight.

After six weeks on the ship, I returned to Wellington and was met by a distant and aloof Rosemary. I'd just finished one of the most inspiring trips of my life to a place that none of our friends had been to. With visions of the aurora and Antarctic wildlife forever imprinted on my brain, I was desperate to share my experiences with her. But she seemed aloof and cold, with little interest in me or what I wanted to tell her. After a heated row a couple of nights later, she told me that she wanted to explore her 'female side' and that I was not part of that equation. It was clear that she felt our marriage was over. Our relationship had been deteriorating for some time and I'd had a premonition that this was coming, but fear and dread had caused me to bury the truth. Despite everything, we'd had some wonderful times together and I had been clinging to the hope we could make the relationship work. But Rosemary made it clear there was nothing I could do. When the inevitable came, I was absolutely devastated. I did not want it to end, especially in this way.

In the months that followed, with time for self-recrimination, I realised that I had been an immature idiot and had not paid much attention to Rosemary. We had been married for a little over four years and, almost from the beginning of our relationship, I'd put a huge amount of time into my own work ignoring her own work and ambitions. Three years after our separation she began a PhD in education in the US, and has since had a very productive career there in academia, where she is well respected by her students and colleagues.

For my own sanity as well as hers, I'd decided to move out of the house in a separation that I'd hoped would be temporary. And so began one of the darkest years of my life. I wandered through 1974 like a zombie. I could not face trying to impose on friends. After yet another furious argument with Rosemary, I'd thrown a few belongings into my old car, driven off and vaguely thought about trying to sleep in it that night somewhere on a quiet street in Lower Hutt. However, for some reason I remembered an older couple, Keith and Mary Gibson, who also lived in Lower Hutt. One of the young lawyers in Keith Gibson's firm had sorted the legal paperwork for the house Rosemary had bought us while I was working at Scripps. I'd only met Keith and Mary once before, but somehow their kindness stuck in my memory. I reasoned that they did not know anything about Rosemary and me, and would be impartial about our situation. Perhaps they could put me up for the night and give some advice? I pulled up outside their place and sat in the car in the darkness for about an hour before plucking up the courage to knock on their front door. A surprised Mary took one look at me and said, 'You'd better come in.' I stumbled inside and burst into tears.

During the next few hours, Keith and Mary listened without judgement as I poured out my pain. I was at the end of my tether, physically and mentally. The effects of the breakup were exacerbated by eighteen months of pent-up stress from keeping Baring Head going. I was young and fit but very sensitive; my heart was broken and the double-edged stress was more than I could handle. I knew I needed help, and somehow that night I had blindly stumbled into a situation where help was forthcoming. This kind elderly couple gave me a bed for the night, and during the next week they helped me face the stark reality that my marriage was over. They arranged for me to see both a marriage guidance counsellor and a medical doctor. Both of them were frank and advised that no matter how hard I tried, there was no way I could put my marriage back together again. And if I kept on with the long hours and stress at Baring Head, I was going to be in even more trouble physically. It became clear that I was suffering a serious breakdown caused by overwork, a broken heart and emotional stress. Keith and Mary offered me a room to stay, and together they nursed me back from the brink.

Over the next couple of weeks, I took the first sick leave I'd ever taken, spending hours in bed at Keith and Mary's place sleeping and resting. My physical strength and fitness returned, but my mind was in a turmoil trying to come to terms with my loss. Many people compare a marriage breakup to a death in your close family, and it was like that. I felt a huge loss inside as though part of me was missing, a gnawing feeling that persisted almost every moment of every day for months.

After I'd been living with Keith and Mary for a couple of months, Athol Rafter asked to see me. I hadn't wanted to bother anyone with my personal problems, but it seemed people had taken notice, and Athol said he was sorry to hear about my marriage. There was no getting round it – he told me I needed to take a break.

'You've got your whole life to be a successful scientist. How about I give you six months leave without pay and you bugger off and sort yourself out? When you come back, I'll give you your job back.'

My first thoughts were of Baring Head. While Owen Rowse had done a great job running the station when I was in Antarctica, he would need more training. I promised Athol I would think about it, and that night had a long talk with Keith and Mary about the offer. I tried to think through what Athol had said. But I was exhausted and fragile; my self-confidence had taken a big hit. I came to a decision. I agreed to take up Athol's offer in a few months, which would give me time to regain my confidence and strength as well as train Owen further.

For those first couple of months with Keith and Mary, I stayed in their home after work, chatting to them as well as reading and watching TV. I had no strength for anything else. I would find myself breaking down and weeping at unexpected moments. I kept mostly to myself but, somehow, Keith and Mary were always in the background supporting me. Socialising was out of the question. I just did not have the strength to rebuild my social life or try to meet new people. A big problem was that most of my friends were also Rosemary's friends. It was not a matter of taking sides, but many of our friends seemed to be uncomfortable when I contacted them. I'd read about divorcees being a perceived threat to other people's relationships and I'm pretty sure that I experienced that. I felt isolated – in many ways it was as though Rosemary had died; a huge part of my life was black and empty. Keith and Mary also warned me about advances from other married women. I did experience this, and after my own problems there was no way I wanted to compromise someone else's marriage.

Keith and Mary expected nothing in return for taking me into their home. Without their help, love and persistence I would have crashed and burned. Somehow, and with their unrelenting support, I made it through that year. In the eight months I'd been with them we formed an unbreakable bond – they were childless, and treated me as an adopted son. I will never forget their love and care.

During this period, Dad had a heart attack while he and Mum were travelling in England. Seeing his white, exhausted face when he arrived back at Wellington Airport was a terrible shock, and I resolved not to burden them with my own problems. Mum and Dad had a strong church background and my separation did not sit well with them, especially Dad. They were relieved that I was well taken care of by Keith and Mary, and after a week in Wellington they returned to New Plymouth.

Gradually the months of 1974 ticked past and I got used to the comforting routine of life with Keith and Mary, helping them with chores and easing back into full-time work at the INS and Baring Head. I focused on improving my mental and physical health. It took many months, but gradually my pain and grief began to subside as I adjusted to being alone.

CHAPTER 7

THE WORLD EXPERTS MEETING

By October 1974 the CO_2 concentration at Baring Head had reached about 326 ppm, another 1 ppm up on what Peter and I had observed the previous year. This could only mean the CO_2 concentration of the whole atmosphere had increased by that amount due to the combustion of fossil fuels. Keeling and Revelle had surmised that the southern oceans would be a major sink for excess atmospheric CO_2, but the Baring Head measurements were consistent with the fact that only some of it was being absorbed – the rest remained in the atmosphere, adding to the earth's greenhouse effect. That year, when I heard about the plans of the 'seven sisters' oil companies to increase exploration due to demand, with a feeling of dread I knew that the madness would continue. They were becoming the wealthiest transnational companies on the planet. There was no way they would curb their profits in the name of our research.

By this point I had been working on atmospheric CO_2 for well over four years and, with the exception of Keeling's group at Scripps, I had more experience with the equipment and calibration techniques needed to make these measurements than almost anyone else in the world. I was beginning to receive more and more letters from scientists and organisations around the globe wanting to learn from my experience. Labs in the US, Japan, France and Australia – many wanted to use Baring Head as a proven site for making collaborative measurements of atmospheric CO_2, its stable isotopes and other related gases. With more organisations beginning CO_2 measurements, an agreement on calibration standards was becoming urgent so that we could compare worldwide measurements.

Until this time all calibration gases were supplied by Scripps and derived from Keeling's original manometric system. The gases were linked to a sensitive mercury manometer system which measured tiny differences in gas pressures, designed by Keeling in the late 1950s and improved in the 1960s. Using a series of precisely determined volumes, the device, expertly run by Peter Guenther, could provide CO_2 measurements in air to a precision of about 0.2 ppm. This was more than enough to compare CO_2 levels between the two hemispheres and to show the seasonal cycle in the northern hemisphere. But in the southern hemisphere, where gradients between Baring Head and the South Pole and seasonal cycles were less than 1 ppm, the precision of the calibration gases produced in the Scripps lab could be a limiting factor. The demand from other laboratories for calibration gases began overloading the facilities at Scripps, and it was becoming obvious that some kind of international consensus on measurement compatibility was needed.

With this issue in mind, Keeling approached the World Meteorological Organization (WMO), a United Nations body based in Geneva, for help. The WMO was fully aware of the potential effects of increasing atmospheric CO_2 on climate, and in a remarkably short time they organised the first 'World Experts Meeting on Carbon Dioxide Monitoring' to be held at Scripps in March 1975. Keeling invited me to the meeting and asked if I could spend some time there beforehand, working with him on data from the New Zealand programme. The WMO offered to pay travel costs and accommodation for me to attend the meeting, and Athol said he would pay my salary for two weeks in California.

Athol had a further idea. 'Look Dave, why don't you consider this meeting as the start of your six months off? Plan a trip afterwards. You could visit a couple of science labs while away and even think about where you might do a PhD.' When I talked with Keith and Mary they were thrilled. After the dark year I had just been through, they felt everything was starting to fall into place for me and asked where I might like to go after the conference. Without even thinking, I knew the answer. Germany is considered the birthplace of modern atmospheric chemistry. Professor Junge, a world-famous atmospheric chemist, had had a big influence on Keeling's career. I knew I could learn a lot by going there.

'But what about the language?' said Keith. 'You don't speak German.'

'I guess I'll just have to learn,' I said with a grin.

By the end of 1974 I'd started going on dates and to parties with friends, and was feeling a lot better about myself. After months of despair, I was looking forward to the future again. None of the dates I went on were serious; I was looking for companionship and friendship rather than an enduring relationship. And yet I did go out on a number of occasions with Irena Smolnicka, a technician based at a DSIR division in Lower Hutt called 'Geological Survey'. I was fascinated by her: vivacious and outgoing, she was bilingual – a born Kiwi fluent in Polish, her parents' language. The more I spent time with Irena, the more interested I became in her background and thoughts on European countries. She was twenty-two and had decided to travel to Europe on an incredible three-month overland trip from Kathmandu to London, leaving New Zealand in early 1975. The trip would traverse India, Pakistan, Afghanistan, Iran, Turkey and Greece, through what was then Yugoslavia, and north across Western Europe to London. She didn't plan to return for at least a couple of years, getting a job somewhere in France or Germany. To prepare for this she had learned both French and German.

'You know,' I said one morning, as we walked along a street in Wellington, 'I'll be bumming around Europe about the same time as you. Maybe we could meet up?' After my breakup with Rosemary, I was cautious and felt I wasn't ready for another serious relationship. But I'd really enjoyed Irena's company and thought it would be fun to spend some time exploring the continent together, especially as she had some familiarity with French and German.

'Could work,' she said. 'Let's see how we feel in a couple of months.'

I had no clear plan at that stage, but I filed away the thought that I would try and meet her if the opportunity arose.

At Keeling's request I arrived a week early for the WMO meeting in March 1975. Being back in California felt like being home. Everything was familiar, and the friends I'd made during my six months there in 1972 immediately threw a party to welcome me back. I had a wonderful ten days staying with Peter and Chris Guenther in Leucadia, enjoying their relaxed lifestyle, with thoughts of Rosemary far from my mind. Keeling welcomed me back to Scripps, congratulating me on the success of the Baring Head programme and explaining how vital the data would be to justify his new National Science Foundation funding proposal. We then talked for some time about my voyage to Antarctica and I remember asking something which had been bothering me for some time: whether the strength of the Southern Ocean sink might be changing.

'A very important question,' said Keeling. He explained that, if the Southern Ocean's absorption rate for atmospheric CO_2 slowed, we could be in real trouble. The airborne fraction of excess CO_2 in the atmosphere would climb, and more CO_2 in the atmosphere would lead to increasing effects of climate change. I was horrified. It was clear of course that carbon emissions were modifying the atmosphere, and it was already well known that polar regions would change at a rate faster than the rest of the earth's surface. But what Keeling had just told me is what we refer to today as a climate tipping point: where part of the Earth System becomes so disrupted that it creates a climate change feedback loop on a gigantic scale, a dreadful spiral taking us to . . . what and where? In the case of the Southern Ocean, Keeling worried that changes in wind intensity driven by warming atmospheric temperatures could begin to restrict the amount of atmospheric CO_2 absorbed. That something as vast as the Southern Ocean could be changed by humans seemed impossible. There in Keeling's office, I clearly remember a sense of foreboding. Keeling noticed the look on my face.

'You can see how important good sites like Baring Head will be to figure this out,' he said. 'We're going to need a lot more baseline data from the Southern Ocean. It's critical that we keep Baring Head running and expand our measurement network in the southern hemisphere.'

The meeting of CO_2 experts was held in a small cottage on a hillside above Scripps, overlooking the ocean. There were fourteen delegates: seven from the US and the remainder from New Zealand, Canada, France, Australia and Sweden, with a representative from the WMO in Switzerland running the meeting. I was the youngest person there, but due to my work at Makara and Baring Head in New Zealand, the person with the most experience operating remote atmospheric CO_2 monitoring equipment. Forty-five years later I am one of the last surviving attendees of what turned out to be a historic and groundbreaking meeting, the first of what is now a biennial international event involving hundreds of atmospheric scientists from around the world.

As a young scientist, playing an important role at that meeting was one of the highlights of my career. I could talk freely and in detail with other delegates of my experiences and the equipment and techniques I had developed, and share recommendations. Everyone at the meeting was concerned about the increase in atmospheric CO_2, and a lot of discussion centred on making atmospheric data available to the public and increasing public awareness. In the mid-1970s, only those of us involved with atmospheric measurements as well as climate modellers knew about the dangers we faced. But there was still no public awareness of the coming crisis, and most scientists working in other disciplines had no knowledge of the issue either.

With so many organisations interested in making atmospheric CO_2 measurements, it was essential that the data collected in different locations could be compared on the same calibration scale. The need for a central calibration laboratory to prepare calibration gases for use at CO_2 monitoring sites around the world rapidly became a central theme of the conference. Because Scripps had been preparing calibration gas standards for more than fifteen years, it was decided that in the interim they would take over the supply of gases to all contributing international laboratories; Scripps would be called the WMO Central Carbon Dioxide Calibration Facility. Funding would come from the WMO and the US National Science Foundation.

The arrangement would be taxing on Keeling's operation, as there simply were not enough skilled people available. Two delegates at the meeting, from the US National Oceanic and Atmospheric Administration (NOAA), wanted to help. This marked the beginning of a robust partnership; during the next few decades NOAA, based in Boulder, Colorado, went on to take a leading role in the science of atmospheric trace gas measurements including CO_2, as well as offer logistical and technical support for the production of CO_2 calibration gases used all over the world.

There were many other issues discussed at the meeting, including the accuracy of measurements and the carrier gas effect I'd been the first to see about three years before, which had now become a problem requiring international cooperation to solve. The meeting was stimulating and exhilarating; as a junior scientist I felt humbled to have been invited to join the world's leading atmospheric CO_2 experts.

After the meeting, I spent a couple of days at Scripps working with Peter Guenther and others on preparing the Baring Head data for the funding proposal. However, it was clear there were some calibration issues that would need a lot of work to correct. It would involve deriving some mathematical functions to fit the data, and a lot of computing. The day before I was due to begin my six months' leave in Europe, Keeling called me into his office and asked me to stay on to correct the data.

I felt a knot in my stomach. This was a true conflict. On the one hand I was totally dedicated to doing the best job I could on this important project. But on the other, I had burned myself out keeping the project running at Baring Head, and it had cost me my marriage. I needed a clean break to clear my mind and reflect on what mattered to me. If I stayed, I would pay for it physically and emotionally. It could also compromise the future of my scientific career.

This was a huge moment – it was an honour to be asked to stay, but though Keeling had been a major force and mentor in my career to that point, I knew a life in science ahead of me would involve expanding my horizons. Athol had already alluded to this when we'd discussed the possibility of me doing a PhD in atmospheric chemistry. All this flashed across my mind as Keeling waited for me to respond. I thanked Keeling and apologised, saying that I couldn't stay for personal reasons. Instead, I asked if I could return to work at Scripps the following year. Keeling assured me I was welcome any time, and wished me luck on my travels. The next day, I said goodbye to Peter and Chris and boarded a Greyhound bus in San Diego.

PART III

1975-1980

330-337 ppm

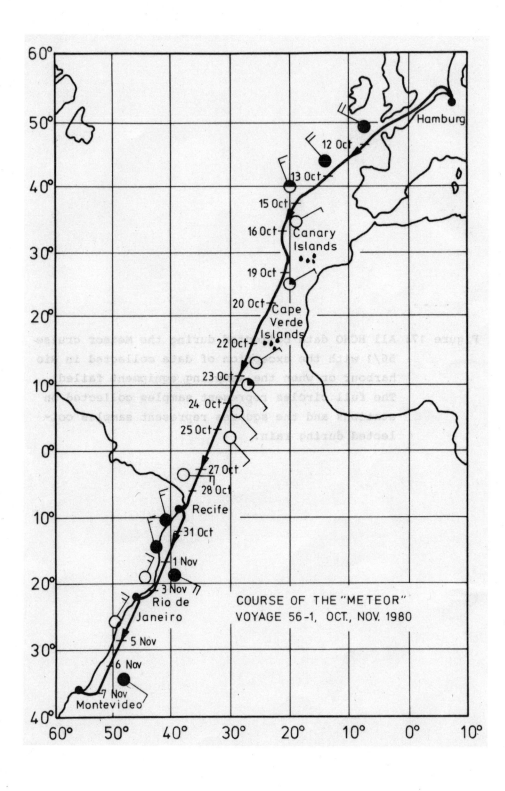

COURSE OF THE "METEOR"
VOYAGE 56-1, OCT., NOV. 1980

CHAPTER 8

MEANDERING INTO THE LIGHT

Back in the 1970s, the traditional Kiwi overseas experience involved joining the quarter of a million or so New Zealanders who lived in London at any one time. Leaving our small islands and bumming around Europe was considered a rite of passage for those in their late teens or twenties. Many travelled by ship, until in the mid-1970s flights to London became cheaper and were the preferred option. What I was doing was unusual, flying to California then travelling overland to Calgary in Alberta and on to London. But really it was the continent that called me – its history, science, languages and culture.

For the first time in many years I had a blank slate in front of me. I had a couple of stops planned at labs in the US and Canada before I flew out to Europe but, for most of the time, I was just going to do what felt right. Six months off to discover the world and put my broken pieces back together.

I spent the next three weeks meandering up the west coast of the USA by bus through California, Oregon and Washington to Vancouver in British Columbia. Everywhere, I immersed myself in the culture and met people who almost without exception I found generous, interesting and friendly.

In Canada I decided to meet my uncle Charlie who lived in Lethbridge, about an hour south of Calgary. It was the end of March, incredibly cold, with frozen lakes and snow everywhere. The Bow River in Calgary was frozen solid, the ice metres thick. The few days I spent with my uncle were the coldest in my whole life and I had not brought the right clothes to cope.

Uncle Charlie had emigrated from England to Canada almost forty years before and had worked as a cook for lumber gangs in the far north. I was fascinated by his stories of survival. He and my dad got on well because they were the only two of four siblings who had left England and had an appreciation of the world beyond. I was going to visit another uncle and aunt in Essex, England, in the next couple of weeks. Uncle Charlie gave me an appraising look and warned me that the visit 'might not go as I expected'.

Through Uncle Charlie I learned about the Hutterites, who had colonies around Lethbridge and many other parts of Alberta. The Hutterites are conservative and closed self-sufficient communities who scarcely mingle with other Canadians, but are respected for their hard and fair work ethic as well as their strong religious beliefs. The Hutterites originated in Southern Germany and many migrated to the US in the late nineteenth century. After persecution in the US over their pacifism during World War One, many of them moved to Alberta and Saskatchewan. I was fascinated by their culture and asked Uncle Charlie if we could go and visit some. He looked dubious but took me to the nearest Hutterite settlement, about 10 kilometres outside Lethbridge. I blithely went and knocked on the first door I could find and when somebody opened it, I smiled.

'G'day. I'm from New Zealand. I've heard a lot about you and it's great to meet you,' I raved.

'Huh!' was the astonished response. The door closed gently but firmly in my face.

Uncle Charlie laughed and shook his head. I knew he was thinking, 'Dumb Kiwi!'

A week later I was in Thaxted, Essex, marvelling at the thatched cottages of the village in which my father was born. I'd flown into London a few days earlier and spent a couple of nights in a ramshackle hotel before taking a train to Essex to meet my uncle and aunt.

In Thaxted's main street, outside a shop with a sign that read 'Lowe and Co, Corn and Coal', stood an elderly man who looked a lot like my father. He welcomed me inside. The shop was owned by

Uncle Harry and had been in the family for generations. He, Auntie Greta and my cousin Evan lived next to the shop; parts of the house had been built more than 300 years ago. They made me feel very welcome. This was the first visit they had had from a nephew living on the other side of the world. Naturally I was curious to learn about them, but I soon discovered that communication was a problem. My uncle and aunt had never been to London, less than 100 kilometres away, let alone overseas. Despite being very hospitable and friendly, they had no interest in my studies and work, or what I thought about the world and climate change. We talked about immediate family matters, the corn and coal business, and the local church. That was fine; I soon learned to be patient and limit my questions to issues they were comfortable with. They'd heard about my marriage breakup and this really disturbed them. Auntie Greta in particular disapproved of my situation.

'You should've handled things more properly,' I remember her saying. 'Perhaps you should go back and live with your parents rather than gallivanting around the world!'

It was an odd feeling to be in a tiny English village where generations of my father's family had lived, married, had children and died. The first Lowe to live in Thaxted was a flour miller, Robert Lowe, who came from a village near Cambridge, where he'd worked on a watermill in the early 1800s. In Thaxted he worked in a windmill which is still standing today on top of a small hill nearby, faithfully restored and looked after by a local historical society. I've often wondered whether having a blood relative who worked in a windmill almost 200 years ago has played a role in my fascination with renewable energy.

My father had been born in the house where my uncle and his family lived. I felt that I should feel part of the family and village, yet somehow I did not. I thought of myself as a New Zealander. The physical similarity between my father and Uncle Harry was remarkable, and I looked like my first cousin Evan of about the same age. But that's where it ended. My background was different, and I never seemed able to communicate thoughts and feelings that were important to me. In spite of their open, hospitable nature, I felt sad and lonely. I left wondering what I could have learned if we had found a way of talking to each other comfortably.

Afterwards, I discovered many other young New Zealanders experienced the same feelings when they tried to meet relatives in England. Was it a matter of unrealistic expectations? Should a young New Zealander with different interests and a different cultural outlook on life expect a relative in England to understand them? Oddly, I'd been able to share and communicate more with the strangers I'd met on my journey through the US and Canada. I pondered this as I continued my travels, catching a ferry from Dover to Calais in France.

Apart from a few days in Mexico when I'd been working at Scripps, France was my first experience of being in a country where English was not the main language. From the beginning of each day to the end, I was fascinated by trying to make sense of the French signs, newspapers and books, and speaking to as many people as I could, using my high-school French and a simple phrasebook. Paris was a magnet for young people – hanging out at a street café with someone you scarcely knew, watching the people walk by and listening to snatches of conversation, created an intoxicating feeling of being a part of the city. It was in Paris that I first discovered art galleries. I was in awe at how these paintings spoke to me and touched my soul, despite my having grown up on the other side of the world. But after a week in Paris I began to feel restless; I had a three-month Eurail pass in my pocket, allowing me full use of trains anywhere in Europe. Where should I go next? Someone at a street café suggested Spain and, on a whim, I jumped on the next train to Madrid.

So passed weeks in a mad blur of hitchhiking, camping and sailing around Spain, France and Italy. As I travelled by ferry to Greece, I recalled Irena Smolnicka, the young New Zealander with Polish parents who was currently making her way overland from Kathmandu to Europe. I thought it would be fun to try and touch base with her as her bus passed through Athens. We had a lot in common and she'd interested me more than any other young woman since my break-up with Rosemary more than a year ago. If I made it in time, I'd probably be able to meet her group when they arrived at their hotel. My Eurail pass could take me most of the way there, and if I missed her there would probably be another chance. I still had five months before I planned to head home.

Now barely awake, I looked out at the coast of Albania to the east as the ferry lumbered its way towards Greece. Around me, lying on the deck, were about thirty other young people all bound for the Greek island of Corfu, the ferry's next stop. It was getting lighter. I stood up, stretched, and watched the coast of Albania slipping past as the sun rose over barren rocky hills. I felt an inner peace. My travels were working, just as Athol Rafter and Keith and Mary had predicted they would. I was beginning to heal from the ravages of overwork and the breakdown of my marriage.

Athens was charming, shabby, intense, crowded, dirty, fascinating – and beginning to recover from a recent military coup. I loved it. I spent a few days in a hostel full of young German students and wandered aimlessly through crowded markets and plazas. Soon I was waiting outside the hotel where Irena's bus was due to turn up. What would she think? The long-haired, screwed-up guy she had gone out with a few times, turning up in Athens. Right on time, a dusty Penn Overland bus pulled up and young people spilled out.

'What are you doing here?' yelled an astonished Irena, as she gave me a hug.

'Oh, I just thought it would be interesting to see how your trip's going.'

'Gosh, where to begin?' she laughed. 'Let's get a drink.'

In a quiet bar I listened to her stories of an amazing three-month journey from Nepal through Asia. She'd crossed through countries like Afghanistan, Pakistan and Iran. I felt her joy – the trip had changed her outlook on life, and she was eager for more adventure. I spent the next three days with her exploring Athens and feeling wonderful.

But there was an unspoken reserve and a wariness between us. Irena was full of new experiences and had a taste for adventure. She had lived her entire life at home in her close-knit Polish family and this trip was the first time she had lived away from them. She needed to continue her travels and discover what was important to her. Meanwhile I was in another phase of my life, enjoying myself but still coming to terms with the previous terrible year. We wanted different things and needed to go in different directions. We agreed that, depending how each of us felt, we might meet up in London in a few months' time. I gave her an address in Brussels where I planned to stay with my musician friend from university, as well as the address of my aunt and uncle in Thaxted. She told me she would drop me a line in a couple of months and let me know how things were going. The next day I caught a ferry back to Italy and spent two days in trains travelling north to Germany, as she continued her overland bus trip through what was then Yugoslavia and on to London.

I spent my first few days in Germany in Tübingen, a picturesque university town not far south of Stuttgart. From the first moment the country captivated me; at that stage I had only a basic knowledge of German, but I loved the sound and feel of the language and resolved to try and learn it. The narrow winding streets of Tübingen along the Neckar River were lined with small cafés and bars full of students. In no time I'd made contact with several science students as well as a local named Sabine, a friend I'd made in Spain some weeks earlier when we were thrown off a ship from Algeciras to Tangiers. Through them I learned about the German education system, which offered free tuition for foreign students. For years the government had taken the view that supporting foreign students would be a good investment, with many talented young people choosing to stay in Germany and thereby boosting the economy. In addition, the government invested heavily in scientific research and development including basic research like climate science.

The more I learned, the more excited I became. I had had a lot of experience in my field, but to progress my career in climate change I needed to study for a PhD in a subject like atmospheric chemistry. Courses like this were not offered at the university in Tübingen, but the students I met assured me that world-class studies in atmospheric science were offered elsewhere in Germany. I knew that Keeling had been inspired by Professor Christian Junge at the Max Planck Institute for Chemistry in Mainz. Junge is generally credited as the father of 'air chemistry' – I even remembered having read a book of his, *Air Chemistry and Radioactivity*. I filed away ideas of study and thought about it on trains to the north of Norway.

After travelling for a month through Denmark, Sweden, Norway and northern Germany, I decided I needed a break to catch up with my musician friend, Allan James. He and his wife, Jan, and ten-year-old daughter, Teresa, lived in a decrepit downstairs flat in the heart of Brussels. There were three more flats upstairs and the whole building shook as ancient trams rattled through a major intersection a couple of metres from the front door. A shabby, dimly lit corridor served as an entrance to all the flats. There was no bathroom in Allan and Jan's flat; they had to share one at the top of a rickety staircase with another flat on the second floor. The occupants of this flat, whom I hardly saw, used to routinely flush their dog's droppings down the toilet in the shared bathroom. The place stank. The stench hit you halfway up the stairs and left you gagging.

But the lifestyle was captivating: Allan and Jan introduced me to a lot of their Belgian friends and made me welcome. The flat was cramped but they found a place for me to sleep under a double harpsichord that Allan was repairing for a client. Allan used a front room of the flat as a workshop where he repaired various instruments including harpsichords and clavichords. He also taught music theory and performance to local students there. Since leaving New Zealand five years before, he had gained a prestigious qualification in pipe organ performance at a top music school in Brussels and, since then, had scratched out a living repairing antique instruments for friends and clients as well as playing the organ for three different church services. I don't think he earned much teaching music and I couldn't help noticing that all of his students were attractive young women – from the time I'd first met him, Allan had a reputation as a ladies' man.

Allan often used to practise on the pipe organ in the Cathedral of St Michael and St Gudula, built in the 1500s and set in the heart of Brussels. We would climb a winding staircase almost 15 metres above the pews to the main console of the organ, and he would play while I listened in awe. The largest pipes on this organ were 20 metres high and the sound they produced was sub-audible – you felt rather than heard the note as a kind of deep vibration in your whole body. No recording can simulate that sound. You have to be there to feel it.

After months of travel, Brussels was a wonderful place to hang out and take stock. Allan was fascinated by my work with CO_2 and he and his Belgian friends asked me a lot about it. Although they had not been aware of the potential climate problem, they were open to the idea and interested in aspects of the science. Living in crowded Belgium, they had already seen the effects of industrial pollution – such as acid rain, which was dissolving many of the monuments around the city. When I'd raised climate issues with New Zealanders, the response had usually been polite disinterest. This Belgian group felt different – they were an odd bunch of intellectuals with thoughts, observations and queries that often came out of left field. Hanging out with them was a time of catharsis. They encouraged me to open my mind to history, languages, arts, music and different ways of looking at the world around me. Science had dominated my life ever since I had left Taranaki as a teenager. Now, I was beginning to appreciate how knowledge from all fields interacted, leaving me richer and wiser.

Allan and I had been close when we were students together in Wellington, but our bond deepened in Brussels. I spent a lot of time exploring the city and got to know it well. Because Brussels was also a major rail junction, I often used the Eurail pass to do day or two-day trips to neighbouring Germany, Holland, Luxembourg and France.

Allan's attitude to life was irreverent and intoxicating; the time I spent with him has had an indelible impact on me.

Allan and Jan used to throw a lot of parties in their flat, attended by an eclectic mix of musicians, colleagues and others. I would play Allan's guitar and sing while he accompanied me on the harpsichord – a strange combination musically, but it seemed to work. One of their friends was Adele, a young Frenchwoman who had worked as a telephonist in Brussels for ten years. She and I became friendly and, when she had time off work, she showed me around Brussels.

After the parties at Allan and Jan's, Adele would often invite me to walk with her to nearby wine bars. These were intimate cellars open all night, full of odd people, cigarette smoke and music. We would drink rough red wine as she told me about her life in France and Belgium. She seemed wistful. I guessed she had troubles on her mind but she never elaborated and I never probed.

After one late night session drinking cheap wine, she looked at me intently and asked whether she could go back with me to Allan and Jan's flat to spend the night. I was taken aback but really liked her, so agreed. After we'd spent some time together under the harpsichord, there was a thunderous knocking and yelling at the front door of the building. I'd been in the flat for about a month and was used to the constant noise outside, so I just ignored the commotion. But the banging and shouting continued and eventually Adele got up.

'I think that's for me,' she said.

She went to the front door and opened it. By this time, Allan and Jan had got up and the three of us stood inside watching, incredulous. Adele and a large man with a wild look in his eyes were toppling over the front steps out into the street, rolling around on the cobblestones punching each other. This was accompanied by a torrent of yelling, expletives and abuse in French. It was 3am. There were no trams running but a maintenance gang was working on the tram tracks. They shut down their welding gear and came to watch the fight, which was getting nasty. After a few minutes of brawling the big man got up, looked at us three New Zealanders standing in the doorway, and glared at me. He and Adele staggered off together into the night.

'Who was that?' I asked.

'Oh, sorry about that,' said Allan. 'I forgot to tell you about Adele's Belgian boyfriend, Georges. He seems a bit jealous, don't you think?'

One glance from the Belgian giant had told me all I needed to know. A few hours later I left Brussels for the Channel ferry, and for England.

It was almost four months since I was last in Thaxted, and a number of letters were waiting for me – from my parents, Athol Rafter, Irena and other friends. Athol wrote with news of Baring Head; Owen Rowse had managed to keep things going but it was urgent that I get back to take care of several difficulties. In the meantime, the CO_2 measured at Baring Head had gone up another 1 ppm compared to the year before. And Athol wrote that, in addition to sorting out the Baring Head issues, Keeling wanted me back at Scripps as soon as possible. The increase in CO_2 had never been far from my mind when I was travelling, and I felt a responsibility to get back to work soon.

In the meantime, I had another three weeks' leave and planned to enjoy them. Irena had written suggesting a couple of weeks hitchhiking around England and Wales, dropping in on people she knew through her family. I jumped at the chance.

Together we hitchhiked from the outskirts of London with a small tent, camping on towpaths beside canals and in camping grounds in the Lake District, and staying with friends of her parents. In Wales we struck freezing rain while we were hitching through a high inland pass. Fortunately, a ride in a Welsh blood laboratory van full of fresh blood samples took us through to a tiny Welsh village beside the Irish Sea where it was much warmer. After all I had seen and experienced during the previous months and now, on this relaxed trip with Irena, I felt a peace and fulfilment that had eluded me for many years. She was a wonderful, undemanding companion, interested in everything we saw. The wariness and reserve that we had felt in Athens a few months before was gone. Neither felt any permanent commitment to the other, but there was a bond and trust growing between us. Yes, we still wanted different things: she planned on staying in Europe to work, with no fixed idea of when she might return. And I was committed to getting back to the CO_2 measurements in New Zealand and work with Keeling at Scripps.

A week later, Irena and I were standing on a train platform in Cologne. It was a familiar place, an enormous rail hub I'd passed through several times over the previous months. The noise under the station's main canopy can be deafening, with trains pulling in and leaving every couple of minutes at around twenty different platforms.

Irena was bound for Warsaw, where she would visit relatives for the first time. The week before, in Wales, I'd decided to travel back to the continent just to keep her company at the start of what was going to be an epic trip across Russian-dominated East Germany and Poland, something she was very nervous about.

On the platform waiting for her train we looked at each other, acknowledging an unspoken and growing bond between us. But Irena's journey on the continent was beginning, and my responsibilities with the atmospheric CO_2 programme called me home to the other side of the world. We said our goodbyes, trying not to address the uncertainty of our relationship ahead. Irena gave me a hug. 'Perhaps we'll meet again somewhere?' And with that she climbed aboard the train. I waved goodbye, watching it speed off towards the east.

A few days later I looked down over Germany from 35,000 feet. In less than two days I would be meeting my parents, friends, and Keith and Mary back in New Zealand. I felt like a different person. Europe had repaired my mind and body and allowed me to think about what I wanted out of life. I'll never forget Athol Rafter's kindness, giving me six months off work to find myself – I'd done that and looked forward to getting on with what life had in store.

CHAPTER 9

AROUND THE WORLD
TO CALIFORNIA

I was back at Baring Head, spending days and nights with the constant wind my only companion. The noise was often a deafening whine modulated by an eerie shrieking. One superstitious lighthouse keeper suggested it was the ghosts of drowned sailors shipwrecked on the rocks below. I'm not superstitious, but on my own at night with the wind wailing, I would catch myself looking over my shoulder. As the wind screamed over the equipment building, an irregular vibration frequently started, with the air pressure changing rapidly inside due to some kind of vortex effect. The impact on my ears and body was something I had to get used to. I was fit and young back then but that environment could wear me down. I'd return home at night exhausted, my mind numb from thinking through myriad equipment problems compounded by the physical battering of the wind at the site and the corrugations of the dusty dirt road. Nowadays laboratory health and safety are emphasised. I wonder what current-day safety officers would think about someone transporting liquid nitrogen and high-pressure calibration gas cylinders to an isolated sampling station in a decrepit old car.

When I'd arrived back from Europe towards the end of 1975, the atmospheric CO_2 concentration at Baring Head had climbed to 327.5 ppm, more than 5 ppm higher than the levels I had first measured at Makara five years before. Somehow my belief in the importance of the measurements I was making kept me going. I was pouring all my energies into the project. Years before, I had dreamed about doing something like this with my life. The work was demanding but it was what I really wanted, and despite the adversity and endless equipment problems, I felt I was making a difference.

Almost from day one of my return, the equipment at Baring Head started to play up. It had been running well and Owen Rowse had done a great job keeping the measurements going in my absence. But now, almost every part of the analytical set-up seemed to delight in throwing problems at me. Worst of all was the URAS-1 analyser which I had been using since the beginning. In 1970 it was already an old piece of gear, designed to measure methane and CO_2 gases in coalmines in Germany and modified by Keeling's group to work with atmospheric CO_2. The electronics were designed around vacuum tubes, many of which had been developed during World War Two and were constantly failing. It was getting harder to source replacements. In the mid-1970s the solid-state revolution was underway and equipment was being designed around stable and reliable transistor and integrated circuit technology. But in the meantime, I had to persevere with the old tube-based analyser.

There were many other vexations. Among the worst was the state of the air-lines I'd been using to collect air from the top of the mast at the edge of the Baring Head cliff. When I'd first started, I used high-purity polyethylene tubes which I fed through garden hoses for protection before hoisting them up the mast using a rope lanyard. This worked well early on, but the constant battering by gale-force winds meant that I had to renew them every few months. They were also covered with salt. Eventually I had to replace everything with marine-grade stainless-steel tubes bolted to the mast with stainless-steel fittings. This was an expensive solution but it handled the conditions well. But changing to new air-lines required exhaustive parallel testing to make sure that the data had not been compromised.

I was always busy, and the pattern was always the same: tackle a problem, solve it and move on to the next one. Work and sleep – sleep and work – so the weeks and months passed.

Since my return, I'd been getting frequent letters from Irena in Europe. Her trip to Poland was a mixed success. Although the circumstances were different, she'd experienced similar issues with her Polish relatives to those I'd had in Thaxted. She found her Western upbringing completely different to that of her cousins and aunts and uncles, who were living under a repressive communist regime. She'd always been proud of her Polish heritage, but once there she felt lonely. Eventually she returned to Western Europe and travelled through Germany and France looking for work. She found a fabulous job at a freshwater research laboratory in France on the shore of Lake Geneva, not far from the Swiss border and the city of Geneva. I loved hearing from her and wrote back as often as I could. She often mentioned our hitchhiking trip around England and Wales. My heart warmed when I thought about how fun and how natural it had been to travel and share experiences with her. Now we were locked into jobs on opposite sides of the world.

Mum and Dad had been thrilled to see me home from Europe. Dad was fascinated by my accounts of the visits to his brothers, Charlie in Canada and Harry in England. He was a little upset when I told him about my reception in Thaxted but agreed that 'the folk there were different'. He had left home at sixteen and gone to sea as a radio operator. He'd never lived in Thaxted again.

Sadly, the wonderful relationship I'd had with Dad when I was young had deteriorated. He was disappointed about the breakup of my marriage and disapproving of the new life I was trying to find, hanging out with people he thought were a bad influence. He was quite religious and played the organ at a church in New Plymouth. On one occasion when he was upset with me, he said, 'Dave, you need to get back to the altar!' He was still weak from his heart attack the year before but, instead of compassion and understanding, I missed the opportunity to show my regard for him.

Tenants were living in the house that Rosemary and I owned in Lower Hutt. Rosemary wanted her share of the money from the house. Rather than sell it, I was advised by Athol Rafter and others to keep ownership, and had had to take out another mortgage so I could pay her out. This left me paying four mortgages, and I had to put tenants in to help with the financial strain of the payments. The situation felt quite unfair. I'd supported Rosemary through her last years at university and teacher training, and had put most of the money into the house. But I agreed to the suggestion by her lawyer of a 50/50 split of our assets, and I just accepted the decision. After I got over the devastation of the breakup, we parted amicably. I wished Rosemary all the best and we never saw each other again.

I was staying again with Keith and Mary, who continued to treat me like the son they'd never had. They were especially thrilled that I'd made contact with Irena, whom they had met before she left. 'That's an amazing young woman,' Mary observed, 'a lot better than some of the other people you've been hanging out with.' Mary had a direct way of telling you uncomfortable facts.

By mid-1976 I was getting excited phone calls, telegrams and letters from Keeling and Peter Guenther at Scripps. One morning I received a call from Keeling to say that the CO_2 levels at Baring Head were much lower than they were at Mauna Loa, and the same or even lower than at the South Pole.

'So, your idea about the Southern Ocean sink for excess atmospheric CO_2 could be right?'

'Yes, looks that way,' he said. 'It's more important than ever that you keep the record going at Baring Head. But I'm worried about the station's calibration. Would you be able to come over to Scripps for a couple of months to sort out the data?'

Keeling wanted me there as soon as possible. There was considerable interest in our data, and it could be used to support another funding proposal. Athol Rafter agreed I should head to Scripps as soon as possible while Owen Rowse kept the gear running for a couple of months.

Keith was delighted when I told him about the trip, which included a salary as a kind of per diem, but I mentioned that I had to make the travel arrangements myself. He knew I was short of money due to the mortgage repayments and offered to loan me the money for the ticket. At first I was speechless, moved by his incredible generosity. When I thanked him, he seemed lost in thought. Then he looked at me and broke into a grin.

'Think nothing of it, lad,' he said. 'By the way, did you know there are some very cheap specials for "around the world" airline tickets at the moment? For almost the same cost you could go to California via France, for example.'

The penny dropped!

My plane taxied to a halt at Geneva Airport. My ticket with Air France had taken me from New Zealand to New Caledonia and on to Paris, with a connecting flight to Geneva. In three weeks, I would board a flight from London to Los Angeles. But in the meantime, I had a couple of weeks planned with Irena in France.

'Dave, it's so good to see you,' called Irena, running towards me as I entered the terminal. She hugged me fiercely and I knew immediately that my round-the-world trip had been the right decision. This felt so warm and right as we hugged and laughed. She wanted to show me where she worked, so we boarded a small bus which took us across the border from Switzerland to the small French village of Thonon les Bains, on the eastern shore of Lake Geneva.

Irena was working in a nineteenth-century chateau, with water analysis laboratories and offices on the first and second floors and accommodation on the top floor. Hers was a small student room with a dormer window which overlooked the lake. She'd arranged to have two weeks' holiday with me, and we spent the first week exploring a nearby alpine region known as the Haute-Savoie. The French Alps were breathtaking. It was early summer, mostly free of snow and ice, and we spent a lot of time hiking trails near the ski resorts, in awe of the scenery and carpets of alpine flowers.

From the moment I arrived there was a change in our relationship, a kind of shared energy that built as we spent time together. Previously we'd both been reserved, sensitive to each other's needs and without expectations of anything more serious. But now, two years after we first met, things felt so different. Earlier I'd been scared, and had closed my heart to feelings I could not admit to. Now I was opening my heart and daring to love again. One day Irena caught me staring at her in wonder.

'I know what you're thinking,' she said. 'And you know the scientist I'm replacing here at the institute arrives back in about ten days. How about I pack up and tomorrow we take the train for Paris and London?'

'What, just like that?' I said.

'I know I want to be with you,' she said. 'I've missed you.' At this point Irena had been away from home for two years and had started to feel it was time to head back. 'You're going to California in a couple of weeks – do you think I could come?'

The next day we went to the local train station. In fluent French she arranged for her baggage to be sent to London and bought tickets to England, where she would apply for a US visa to join me.

Peter and Chris Guenther were thrilled to hear Irena was coming and invited us to stay with them at their home in Leucadia, north of Scripps. While Irena sorted her tourist visa, I flew out to San Diego.

Soon I was back in one of the laboratories at Scripps working on the Baring Head data, trying to account for a strange calibration issue with the URAS-1 analyser. 'This is so bloody frustrating!' I yelled more than once. Keeling had asked me to work quickly and carefully by hand, despite the fact I could write a computer programme to do the calculations. I was dumbfounded by his insistence I use a calculator, and related the conversation to Peter Guenther.

'Sorry Dave, that's the way it is around here,' he said. 'There have been a couple of major errors in our CO_2 data sets caused by computer programmers not taking account of instrumental problems and inputting incorrect data. Keeling doesn't trust computers to correct data. You're just going to have to get used to the situation and hand-calculate everything. The rest of us have!'

In 1970 I'd already learned how to programme the first computers in New Zealand and had been working with them ever since. I had a lot of experience with formulating physical problems into programmable code which could be used to calculate CO_2 concentrations. Now, with six years' experience programming and using computers in New Zealand and the US, here I was sitting at a desk with a calculator and sheets of paper. It was tedious work and I chafed at it. I groaned, occasionally shouted at the non-responsive walls, and got on with it. In 1976, computer technology was still in its infancy. The large-scale microprocessor chips which form the basis of modern desktop computers had not been developed, and the first IBM personal computer based on the Intel 8088 microprocessor chip was still years away.

Keeling often dropped by the lab with words of encouragement and to see how I was getting on. By this stage I'd developed a polynomial fit – a statistical device – to look at the raw data. We were beginning to resolve fascinating features that we thought were probably caused by large-scale weather events like El Niño systems. El Niño and La Niña are opposite phases of a natural weather system in the Pacific driven by fluctuations in equatorial temperatures between the atmosphere and ocean. During an El Niño event, sea surface temperatures off the coast of South America are unseasonably warm whereas temperatures in the western equatorial Pacific are much cooler than usual. During a La Niña event, the temperature pattern is exactly the opposite. The phenomena have a huge impact on extreme weather events in the Pacific region, including on the generation of tropical cyclones. Warmer ocean water also affects the way that atmospheric CO_2 is absorbed, and it was this connection that Keeling was keen to investigate.

To help run the polynomial fits of the Baring Head CO_2 data, I was using a programmable desk calculator. This allowed me to automate a lot of the mathematical processes, but still required me to be closely involved with the raw data.

'This is great, David,' said Keeling one evening, over a beer and corn chips. 'Are you sure there are no errors?'

'You can never be sure of that,' I said. 'I've just repeated the calculations a couple of ways and the results are the same. Also, last night, during the trip back to Leucadia I figured out something I might have forgotten. This morning I double-checked and the data are fine.'

"So, you did that last night from memory, without writing it down?' he asked.

'Yes,' I said, puzzled.

'That's excellent. It shows that you're really on top of this task – you're feeling it.'

I have clear memories of this snatch of conversation with Keeling and at the time had wondered why he set such store in me being able to remember a detail without having to write it down. Years later I realised it was his own way of working too. He had an extraordinary memory and this, with his attention to detail, contributed to his success. Keeling was often criticised for being pernickety, but it was his very fastidiousness, always doing something the same way until he was satisfied that there were no errors, that led to what is arguably the most important geophysical measurement record of all time – the exponentially rising curve of atmospheric CO_2.

When she arrived, Irena immediately felt at home with Peter and Chris. The four of us went on to build a very close friendship based on common interests including people, languages, the outdoors and the world. I was happy and fulfilled, and Irena felt the same way. Our love and respect for each other was growing and it would be hard to imagine a better place to have spent our first intensive time together than Leucadia, an idyllic town with a small village atmosphere on California's Pacific Coast. Peter and Chris's house was a 1930s bungalow clad with rustic weatherboards and a wooden-shingled roof, set among charming backstreets lined with avocado and jacaranda trees. We even saw New Zealand natives like ngaio, pōhutukawa, flax and hebes planted in people's backyards. The neighbourhood was laidback and friendly. Crime, if you could call it crime, seemed limited to smoking marijuana.

While Peter and I were working at Scripps, Chris delighted in showing Irena the area and introducing her to friends. When the day's work at Scripps had finished, the four of us would settle in to a home-cooked meal then sit outside and drink beer and wine and talk late into the night.

During our stay, Peter and Chris took us out to the Anza-Borrego Desert, about 100 kilometres east of San Diego, for a couple of weekend camping trips. They had a special camping spot called Alder Canyon which could only be reached by four-wheel drive. It was isolated, stark and extreme, but achingly beautiful. 'The desert grabs you,' Peter told us, and it did.

Out in that isolated canyon we would see no one for the entire weekend and marvel at the stars at night. The Milky Way was a bright band overhead with constellations like the Big Dipper and Little Dipper clearly visible. We used to joke with Peter and Chris that, to us Kiwis, the constellations were upside down. During the early mornings and evenings while it was cool, we would take short hikes through old watercourses lined with different kinds of cacti, manzanita and desert sage. Sometimes there would be desert flowers, brilliant reds and tufted yellows. In the evenings we would light a small camp fire and brew sagebrush leaves in hot water, making a bitter, highly aromatic and refreshing drink. Even though it was late summer, temperatures were often more than 40°C during the heat of the day and plunged to 5°C at night.

While researching material for this book, I came across a letter I wrote forty-five years ago to Mum and Dad about camping in the Anza-Borrego Desert. I wrote about how wonderful it was, after being shattered and alone, to be sharing experiences in the desert with someone you love, and that Irena and I were growing closer every day. I'm so glad I shared those feelings with my parents – I know now that it would have been a great solace to them after witnessing the despair I'd been though. And finding those words I wrote to them all those years ago has been a kind of closure for me as well.

*

When I got back to New Zealand after completing the Baring Head calculations, I found that Keeling had sent Athol Rafter a letter about my visit. Excerpts from his letter encapsulate what I had to go through, his expectations for Baring Head and his interpretation of atmospheric carbon dioxide's links to long term global energy supply.

October 18, 1976

Dear Athol:

With David Lowe's return to I.N.S., I would like to convey our very favorable impression of his contribution while visiting us. David's stay was almost entirely devoted to recomputing the results of the field effort at Baring Head for the period December 1972 through December 1975. Not only was this tedious work, but it involved devising means to cope with a time variable non-linear correction to the URAS-1 instrument response at Baring Head. We had no previous experience with this kind of problem, and it proved impossible to set up a computational scheme to deal with it until after we had done considerable preliminary testing of the data. In other words, we had to do some of the computing two or even three times as we gained a better idea of the problem. We were able to give David considerable assistance near the end of his stay, but much of the time he was obliged to carry out the work himself. There were too many hang-ups in the record to consider a computer formulation. We agreed near the beginning of his stay that hand work-up would be preferable. We all hope that this approach can soon be replaced by one less consumptive of labor. David will doubtless discuss these prospects with you. As soon as the typing of the tables is complete, we will send you a draft of the final data report for David to check over. I think that you will be surprised at its complexity and comprehensiveness.

In any case, it is very important that the New Zealand data be published soon. An intense interest is developing internationally to relate the atmospheric CO_2 increase to the long-range problems of energy availability. The next year will include some half dozen symposia and workshops on the CO_2 climate problem. The New Zealand data should be prominent at these meetings.

Thus, I hope very much that David can continue for a time to devote most or all of his effort to atmospheric CO_2 work, and that he can obtain some help on working up the data for 1976. With publication of results the prospects for international funding should improve.

Very best wishes,

Charles David Keeling

The letter encouraged the DSIR to invest further in the Baring Head project, including in the purchase of a new solid state infrared analyser, a URAS-2T. When this arrived, I wasted no time installing it, running it in parallel with the original decrepit URAS-1 at first, to make sure that there was an overlapping series of results. Baring Head was as usual consuming a lot of my efforts but, as the dataset grew, it was becoming more and more valuable. When a time series begins, it's difficult to interpret much from the data. After several years of atmospheric CO_2 measurements at Baring Head, we were able to resolve features that didn't occur every year – for example, those caused by El Niño and La Niña events – and to get a clearer view of the overarching trend.

Towards the end of the year the background CO_2 concentration at Baring Head had climbed to 328.6 ppm, over 1 ppm higher than the year before. When I had first started measuring atmospheric CO_2 the growth rate was almost exactly 1 ppm per year; now, a few years later, the growth rate was over 1 ppm a year. There was still no sign of any related global temperature increase. However, a small but significant increase in ocean acidity had been measured by Peter Guenther at Scripps, as well as at other international labs. This could only have come from an increase in CO_2 being absorbed into the ocean. As far as I know, these were the first indications of the effects of increased atmospheric CO_2; multiple changes were beginning to become obvious all over the planet.

The struggle to keep the equipment running at Baring Head continued. Most of the problems were associated with the worn out URAS-1 infrared analyser running in parallel with the newer URAS-2T analyser. Just as I thought I was getting on top of things, this new exercise created a large amount of extra work. I discussed it with Athol, who shared the good news that the DSIR were now fully behind the INS because they, as well as the New Zealand Meteorological Service, were beginning to recognise the importance of the research. As a result, finally I would get more resources to help run the station as well as assistance with the data analysis and computer programming. By this stage I was getting enquiries from labs in Japan, parts of the US, as well as Europe and Australia about the Baring Head data. One of the more notable requests came from the US Congress, asking about the significance of the southern hemisphere data. I also fielded enquiries from international organisations about running related experiments at the site. Several of these involved measurements like stable isotopic studies which supported and enhanced the value of the atmospheric CO_2 record.

True to Athol's word and within two weeks, I had technical help with the routine visits to Baring Head, and the INS workshop started to repair the concrete building, replacing the floor and installing new cylinder racks. They even built a new access gate with a secure lock. After years of working on my own, I now had company, and I could bounce ideas and problems off other INS staff. Best of all, I no longer needed to spend whole days and nights at the station. The new lighthouse keeper, Steve O'Neill, was popular with the INS staff whom he would take fishing, and he took a keen interest in the scientific work and checking that the equipment was running well. I was mindful that long hours had contributed to my breakdown. Now that I was in a wonderful stable relationship with Irena, I would not compromise it by late and long nights at Baring Head.

Thoughts of going to Germany had never left my mind, but with the constant challenges of keeping Baring Head running I'd always been distracted. Now, thanks to Athol's provision of extra help, my workload had dropped. I found myself thinking again about the idea of doing a PhD. I enrolled in beginner evening German classes and immersed myself in the language.

After persevering for a few months, I found I had learned enough vocabulary to pick up the gist of simple books. This meant I only needed to look up a few words per page to understand the story. This was a breakthrough, speeding up the whole learning process, and after a year I was able to read and enjoy German thrillers and science fiction. The more German I read, the more fascinated I became about the country and its people and culture. However, there were limits to what I could learn at night classes, and I was balancing these with my other commitments. Eventually it was obvious that, to go much further with the language, I was going to have to enroll in a more advanced university German course, probably an introductory one which would allow me to keep working at Baring Head during the day.

It was November 1976 and a cold dread gripped me. My cousin had called to say that my father was seriously ill in Taranaki Base Hospital and that I should get there immediately. My parents had not met Irena, but now they would under terrible circumstances – my father on his deathbed and my mother devastated. As the Austin A40 trundled up the highway to New Plymouth, I wondered how I would cope. But Irena was beside me, and I felt comforted by her support.

We drove straight to the hospital where we were met by my distraught mother. As I held her, she shook uncontrollably and looked up.

'Thank you for making my son so happy,' were her first words to Irena.

In tears, Irena and I followed Mum to a side-room of the hospital where Dad was lying pale and in a coma. Thoughts of what I should have told him and how much I loved him flooded through me, and I was filled with regret and despair. My brother Steve and his

1956. Old New Plymouth airport. A haven for kids at Bell Block Primary School growing up in the Taranaki countryside. (Photo by Bert Lowe)

1958. Unpolluted rivers, streams, fields and bush-clad gullies: an idyllic childhood playground learning to appreciate nature in rural Taranaki. (Photo by Bert Lowe)

1959. Air-traffic radio guidance building on Mt Taranaki, maintained by Bert Lowe for the NZ Department of Civil Aviation. Our family would spend days and often overnight there. Brother Steve, Mum and I having fun in the snow. (Photo by Bert Lowe)

1961. Summit of Mt Taranaki on a summer open climb. Ray Jackson centre, Con Jackson back left, me front left. On the right is Con's sister Belinda and mother Mary. (Courtesy Con Jackson)

1964. Surfing the Taranaki coast as a teenager introduced me to the forces shaping the environment, especially the wind and waves. This led me to study physics at university, an ideal stepping-stone to begin researching atmospheric CO_2. (Photo by Steve Lowe)

1972. Baring Head lighthouse station in the early 1970s. The small concrete building on the right is where Peter Guenther and I set up the equipment to make the first atmospheric CO_2 measurements at Baring Head. The station has since become an essential part of a network revealing a rapidly deteriorating atmosphere.

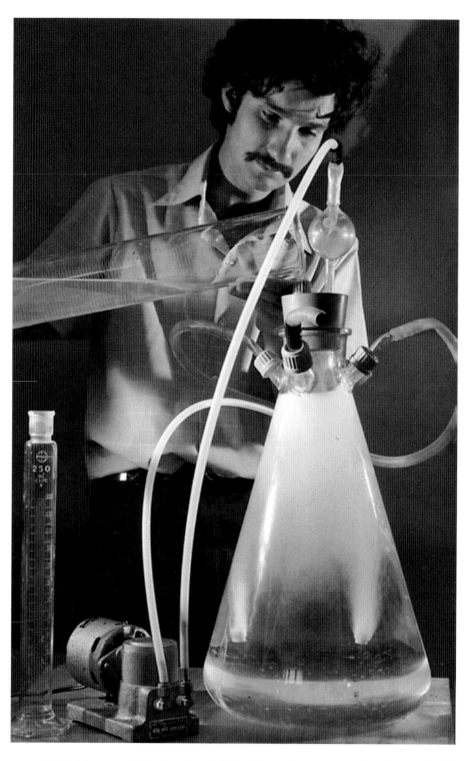

1971. My master's thesis project: lab-based experiments measuring the uptake of CO_2 by seawater.

1971. Old DSIR Shed 2 lab. With
its rats and asbestos in the roof
and mercury on the floor, this
was where I helped develop the
equipment that made the first
continuous atmospheric CO_2
measurements in the mid-latitudes
of the southern hemisphere. The
building is still there, but now it's
a motorcycle repair shop. (Photo
by Peter Guenther)

1972. In this photo I'm setting
up an air-sampling system
using the flagpole at the Baring
Head lighthouse. Note the
regulation INS/DSIR long socks
I am wearing. (Photo by Peter
Guenther)

1973. Peter Guenther and I preparing for a field trip with essential provisions: fish and chips and a bottle of dubious wine. (Photo by Chris Kenyon)

1975. My Austin A40, battered and bruised after countless trips on the rough gravel road to Baring Head. Note the official DSIR lab coat. (Photo by Irena Smolnicka)

1975. First atmospheric CO_2 experts meeting held at Scripps, La Jolla, California. Dave Keeling is second from right, seated; Arnold Bainbridge is back right. I'm standing on the far left, the youngest person at the meeting by far. But after six years at Makara and Baring Head solving equipment problems, I was the attendee with the most experience in making continuous atmospheric CO_2 measurements in remote locations. (Courtesy Peter Guenther)

1975. One of many sessions playing guitar at a party in a student friend's flat in Brussels on my 'healing journey' of discovery in Europe. (Photo by Alan James)

1978. Athol Rafter in his office at INS DSIR. (Photo by Lloyd Homer © GNS Science)

1976. Dave Keeling demonstrating how to take a Keeling flask on the beach in front of Scripps. (© Scripps Institution of Oceanography, University of California)

1976. Being blasted on a typical windy day at Baring Head. Me, Irena Smolnicka, and my brother Steve with his wife and baby. (Photo by Steve O'Neill, Baring Head lighthouse keeper)

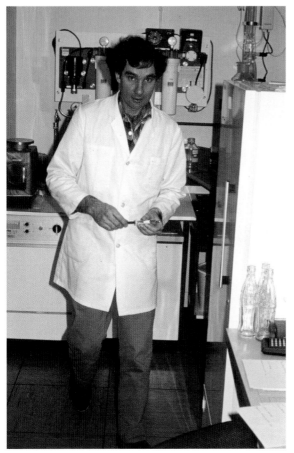

1979. Winter at the Kernforschungsanlage in Jülich, where I worked on my atmospheric chemistry PhD programme for three and a half years.

1979. In the lab at the atmospheric chemistry institute where I developed the equipment for my PhD programme. (Photo by Ulrich Schmidt)

1980. The first major field trip of my PhD programme: the west coast of Ireland, on Loop Head, not far from Kilkee. Ulrich Schmidt making some adjustments.

1980. A lot of my fieldwork involved air sampling from aircraft flying over various parts of Germany. This is one of the research aircraft, a Dornier 28, with its fuselage modified to feed air into my equipment set up in the aircraft's cabin.

1980. Ulrich Schmidt and I were both exhausted with the long days and nights of sampling at Loop Head, Ireland. This is me taking what I would consider to be a well-deserved nap between samples. (Photo by Ulrich Schmidt)

1980. Jülich, West Germany surrounded by sugar beet fields. At the time, the town was ringed by brown coal-fired power plants. The clouds and haze in the picture are partially due to plumes from the plants' giant cooling towers.

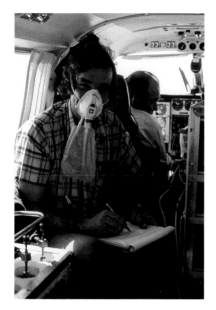

1981. On this research flight I had to wear an oxygen mask due to low cabin atmospheric pressure. During the flight, unwittingly we flew into a 'shooting area' and were buzzed by a NATO jet. (Photo by Ulrich Schmidt)

1981. Carting Greg around on my back in the Austrian alps. (Photo by Irena Lowe)

1981. I'm presented with a 'Doctor's Hat' at my graduation. This is a long-standing tradition in Germany, with the hat being a mock-up of equipment developed during a PhD project. Irena and Greg look on. (Photo by Ulrich Schmidt)

1982. Early days of a giant opencast brown coalmine at Hambach, near Jülich. This mine went on to become one of the biggest in Germany and has been the scene of multiple protests, which are still continuing. Note the giant excavators in the background and the conveyor belt in the foreground, part of the system used to transport coal to power plants up to 20 kilometres away.

1983. Accelerator used for radiocarbon measurements at INS/DSIR in Gracefield, Lower Hutt. (Photo courtesy Gavin Wallace, GNS Science)

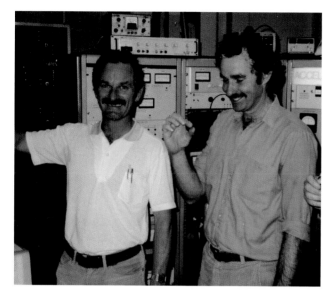

1984. Successful radiocarbon date of an Egyptian mummy. I'm showing Gavin Wallace, on the left, how tiny the sample was. (Photo courtesy Gavin Wallace, GNS Science)

1985. Dave Keeling showing the rise of atmospheric CO_2 on one of the graphs pinned up on a wall outside his office at Scripps.

1987. Peter Guenther in a mellow mood at Scripps.

1989. Stan Tyler, NCAR, sampling air for methane in the Rocky Mountains, Colorado.

1989. Working in the lab at NCAR in Boulder, Colorado. This is equipment I designed to help measure the sources of atmospheric methane. The apparatus is made of glass with cold traps using liquid nitrogen. (Photo by Stan Tyler)

1993. Dave Keeling with his son Ralph at Scripps. Ralph went on to develop a sophisticated technique to measure precise levels of atmospheric oxygen in the atmosphere. He is now a professor at Scripps researching oxygen as well as running his father's CO_2 programme. (© Scripps Institution of Oceanography, University of California, San Diego)

1996. We set up a mass spectrometry lab at NIWA in Wellington in the mid-1990s. Here, I'm showing Martin Manning how the main instrument, a German MAT 252 mass spec, measures samples. (Photo by Gordon Brailsford, NIWA)

1996. In the mid-1990s I set up an air-sampling programme using container ships sailing between New Zealand and the west coast of the USA to collect large air samples in the cylinders I'm carrying. The aim of the experiment was to determine how various atmospheric pollutants are transported from the northern to the southern hemisphere. (Photo by Gordon Brailsford, NIWA)

1996. The TROPAC atmospheric chemistry group at NIWA, Wellington with Nobel Prize-winning visitor Paul Crutzen from Mainz, Germany. (Photo by Martin Manning, NIWA)

1997. This is a later version of the apparatus I designed to make the first measurements of radiocarbon in atmospheric methane. We used these results to show that about a third of the methane in the atmosphere is derived from fossil sources like leaking gas wells and coal mines.

1998. The main atmospheric chemistry lab at NIWA in Wellington. I'd often shout for joy at new results – this looks like one of those occasions! (Photo by Gordon Brailsford, NIWA)

2000. Surfing started my fascination with waves and the environment. Later I took up windsurfing. This is a favourite spot near Plimmerton, not far from Wellington. (Photo by Irena Lowe)

2005. In 2005 I ran an international atmospheric chemistry conference in Christchurch, which attracted 500 scientists from around the world, including a big contingent from Germany. It was a huge job and I could never have achieved it without the assistance of the 'tight five': Bill Allan, Katja Riedel and Anthony Gomez from the TROPAC group, and Richard Richardson, the NIWA accountant. In the front row are the professional organisers we hired from Conference Innovators.

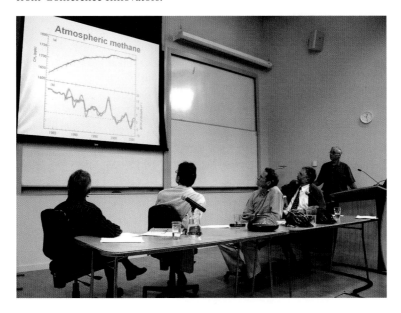

2007. IPCC Summary Lead Authors meeting in Wellington. I'm showing some of the methane results I collated for the award-winning 2007 climate change report. Note the cycle helmets on the podium. I remember a couple of us cycling to the meeting (saving carbon emissions).

2007. Outside the Keeling building at Mauna Loa observatory where Dave Keeling began the first continuous atmospheric CO_2 measurements in 1958.

2007. Peter Guenther at Scripps checking out Dave Keeling's historic mercury manometer. This was used to calibrate the first reference gases for atmospheric CO_2 measurements at Mauna Loa from the 1950s through to the early 2000s at Makara and Baring Head.

2007. Still surfing at the age of 60 at Leucadia Beach, North County, San Diego, but not looking for large waves these days. (Photo by Peter Guenther)

2010. Biking with German friends through a forest of powerlines near Jülich. This small area has more electrical-generating capacity than the whole of New Zealand.

2009. Our family group Suzanne, Greg, Johanna me and Irena. Both Suzanne and Greg were born in Germany and Johanna arrived after we returned to New Zealand. (Photo by Sandra Simon)

2015. I'm flanked by Peter Guenther on the left and Ralph Keeling on the right. This was on the occasion of the Keeling Curve being designated as a National Historic Chemical Landmark at the Scripps building where Dave Keeling set up his atmospheric CO_2 programme. I was invited to speak at the event, one of the proudest moments of my career.

2017. In my role as New Zealand's German Science and Innovation coordinator, I was required to check out German 'Bier und Bratwurst' at a festival in Berlin. (Photo by Rod Harris, NZ ambassador to Germany)

2019. Peter Guenther, Louise Keeling (Dave Keeling's wife) and me outside the Keeling home in Del Mar, southern California. (Photo by Irena Lowe)

2020. Speaking on the steps of NZ parliament buildings at a 'Fridays for the Future' event organised by a group of young climate activists. They have been energised by Greta Thunberg, the Swedish schoolgirl who has inspired millions of young people worldwide in her efforts to bring attention to the urgency of reducing carbon emissions. (Photo by Bill Allan)

wife joined us and we spent that day and most of the following day in Dad's hospital room, sleeping in chairs and on the floor as his condition deteriorated. Once, he started awake and I tried to speak to him as he looked first at Mum. When he saw Irena, he asked who she was and smiled at her, something we will always remember. A few hours later, he quietly passed away.

The funeral passed in a blur of emotion and weariness, and we scattered his ashes in the Garden of Remembrance outside New Plymouth. Irena and I spent a week with Mum, helping her to pick up the pieces. Everywhere we went Mum insisted on introducing Irena as 'my son's de facto wife'. I concluded she thought this was funny, although it annoyed me at the time. By this time, she had formed a bond with Irena, and the fact that we were living together, unmarried, eventually became a non-issue.

Because of the time difference between California and New Zealand, Keeling often used to call me in the morning before I left for work. Usually these phone calls had to do with the practical details of calibration and other issues involved with the equipment at Baring Head. But in mid-1977 I received what seemed to be quite a different phone call.

He was asking me about my personal circumstances, something he had rarely done before. I knew immediately that he needed something from me. I told him I was fine and was getting plenty of help with Baring Head. I waited to hear what was on his mind.

'Look, how would you feel about coming to Scripps for another three months this year so that you can finish computing the Baring Head data and publish it?'

'Three months? I'll need to think about it,' I said. The last thing I wanted was to compromise my relationship with Irena. Of course, I desperately wanted to publish the Baring Head atmospheric CO_2 data – it was time that this early and valuable record appeared in the scientific literature. But I did not want to be apart from Irena; I was conflicted.

After speaking with Irena, we decided to pitch an idea to Keeling. I would work with a DSIR professional programmer, Dave Robinson, to begin working on the data from New Zealand, so would only need six weeks at Scripps instead of three months. Keeling agreed immediately with the revised plan but insisted I start sending examples of the code and output data to Scripps so it could be evaluated and verified by their mathematical modeller and programmer, Bob Bacastow. Over the next couple of months, I began working on the code using a language called PL/1 which had been developed by IBM to handle complex data formats. The language was a so-called high-level language, with the syntax almost like English, and I had to run it on a mainframe computer based in the mathematics department at Victoria University of Wellington. Every couple of weeks I would mail off decks of computer cards and printouts to Scripps. After their initial scepticism, I started getting encouraging letters from Peter Guenther and Keeling, who were impressed with the work so far and wished me and Dave Robinson a 'hearty well done'.

When I arrived at Scripps near the end of 1977, I immediately began working with Bob Bacastow on the PL/1 code for the Baring Head data, and within a week we had it running on a large mainframe computer at the main campus of University of California, San Diego. This was only the beginning of an endless series of batch runs that needed to be done at night, when the computers were available. Initially I'd been living with Peter and Chris Guenther, and the daily commute was an hour each way. To save time, Skip's mother invited me to stay at her place, about a twenty-minute walk from Scripps. I shared a bedroom with Skip's brother, Hayden, but this was frequently embarrassing due to girlfriends who could be in his room at random times – I often had to sleep on the floor of the lounge to avoid disturbing them.

After three weeks, I was absolutely tired out with the routine of running batches at night on the mainframe computer then staggering back to Skip's mother's place, never knowing whether I would be sleeping in a bed or on the lounge floor. After one particularly long night, Keeling called me into his office.

'David, I can see you're making really good progress with the Baring Head data, but I would have thought you would be paying more attention to the outliers and the Kappa correction function,' he said.

A couple of weeks before, we'd agreed that I was going to be working on the outliers later in my visit. But now we seemed to be talking past each other. In a haze of tiredness and frustration, I lost it with him.

'Look, I've been working all bloody night!' I shouted. 'We already agreed that I was going to work on the outliers later in my visit.'

Keeling looked at me, taken aback by my outburst. 'You'd better leave my office and calm down,' he said. 'I know you've been working very hard – I told you that I needed you here for three months, not just six weeks. What you've achieved in such a short time is amazing but you're obviously tired.'

We looked at each other long and hard, and then laughed. I knew that he could be difficult, but I had too much respect for him to be mad for long. I went to the canteen for a large black coffee and a doughnut. That night Hayden didn't have a girlfriend in the bedroom, and I slept like a log after completing the Baring Head batch runs at about midnight.

It took about four weeks of iterative runs on the mainframe to complete all the corrections to the Baring Head data as a series of tables in a data report. The final step was fitting long-term trends and seasonal cycles to establish patterns representative of the Baring Head site. It's hard now to imagine the sheer effort I put in to computing that data set. These days it could be done on a simple desktop or laptop computer, maybe even a smartphone. But I could breathe a sigh of relief that the data was ready for publication. I began writing the first draft of the article.

When I returned to New Zealand in December, Irena and I took a well-earned break tramping together in the Tararua Range and attended a special Polish Christmas celebration with her family, sharing a traditional meal of borsch, pierogi and fish. I loved being part of it, learning about a culture so different to my own, and I could see that her parents' reserve towards me –a young, long-haired, non-Catholic divorcee living with their unmarried daughter – was beginning to evaporate as they welcomed me into their world. After the success of the hard grind I had just completed at Scripps, life felt pretty good.

By the beginning of 1978 Baring Head was running really well, with full support from the INS and Scripps. Soon I was thinking once again about doing a PhD, and when Athol called me in one day for a progress report, he asked if I'd decided where I wanted to study.

I told him that while Professor Christian Junge at the Max Planck Institute was credited as the founder of modern atmospheric chemistry, he was close to retirement. I was also considering a famous institute in a place called Jülich, near Cologne, run by a top atmospheric chemist named Professor Dieter Ehhalt. He'd studied with Professor Junge and had worked with Arnold Bainbridge in the 1960s. Athol approved of Ehhalt — he acknowledged that, despite personal issues, he had the greatest respect for Arnold and suggested I take his recommendation very seriously.

That evening I sent off an enquiry to Professor Ehhalt to see whether he would take me on as a PhD student in atmospheric chemistry. In less than two weeks I received a reply telling me that he was very interested in my background, and could I send him transcripts of my honours and master's degrees. He was also keen to hear about the Baring Head CO_2 programme, so I sent him a draft of the Baring Head article I was working on. Within a few weeks I received the news that the University of Cologne was inviting me to enrol for a PhD provided I passed a German language proficiency test once there. I was offered a scholarship, funded by Dieter Ehhalt's organisation through the German state of Nordrhein-Westfalen.

When I relayed this to Athol, he had even more surprising news. The DSIR had decided to give me an award paying for my studies towards an approved PhD course in Germany. The funds would be paid out via the New Zealand Embassy in Bonn, and they would look after me with medical insurance and help with any other issues. It was obvious that Athol had been working for me behind the scenes and had persuaded the DSIR that the study award would be a good investment for them and New Zealand. All I had to do now, he said, was improve my German. I had been accepted into an extramural first-year German language course at Massey University, one which would involve coursework based on cassette tapes, with supplementary written material. I could complete the assignments at home after work. But there would be one intensive week when I would have to go to Massey University in Palmerston North to pass written and oral tests in a language lab. I would also have to do a final exam.

Athol reassured me to take as much time as I needed. As he had done so many times before, here was this remarkable scientist gently but assertively helping me. I'll never forget his wisdom and generosity.

'What do you think about getting married then?' asked Irena while we were having beer and homemade pizza one evening after work.

'Sounds like a good idea,' I said. 'My mother has been bullying me about it for months.'

We set a date in April 1978 and were married at St Augustine's Anglican Church in Petone, close to where Irena's parents lived. The celebrant's wife was Polish and he used a few phrases in the wedding service which immediately made him a favourite with the many guests from the Wellington Polish community. Keith and Mary Gibson attended and, although he was much older than me, I chose Keith as my best man. Athol Rafter spoke at the wedding and got a lot of laughs when he told the story about me going to work in California via France – the trip where our committed relationship began.

The rest of 1978 passed by in a blur. My paper on the Baring Head project, co-written with Keeling and Peter Guenther, was published in the scientific journal *Tellus*.[9] It was the first of literally hundreds of papers, articles and reports citing a large number of different scientific research projects based at Baring Head.

During 1978 I began to speak out in public about my concerns that CO_2 emissions from fossil fuel combustion were changing the physical and chemical properties of the atmosphere, which could lead to the planet warming. Until this time, I hadn't spoken to media about these concerns, feeling that comments on CO_2 emissions controls should be left to politicians. But as the Baring Head data showed, emissions were continuing to increase. My comments were picked up by Radio New Zealand and newspapers, including a full-page feature in the Christchurch *Star* in which I was quoted predicting that in the next twenty to thirty years increasing CO_2 emissions would become one of the greatest problems facing humankind.

This publicity led to my first unpleasant encounter with climate sceptics, now called climate deniers. I clearly remember trying to discuss the warming potential of increasing atmospheric CO_2 with a couple of aggressive scientists from the Coal Research Association of New Zealand. Neither had any understanding of the physics and chemistry involved and brushed me aside. Their message was simple: 'This is hype and hysteria – the earth is not warming.' Daniel Patrick Moynihan, a well-known American politician and diplomat, is quoted as saying, 'Everyone is entitled to his own opinion, but not his own facts.' To manipulate or falsify verifiable data records and the science underpinning them, in an attempt to confuse the public and deny a climate emergency, is in my opinion a crime against humanity. The malicious manipulation and trivialisation of scientific data by climate deniers is a cross that myself and other climate scientists have had to bear for decades.

Before I knew it, I was completing the handover procedures for Baring Head and passing the torch to Martin Manning, a brilliant theoretical physicist who, in particular, greatly improved the measurement system by introducing a rigorous calibration procedure based on multiple reference gases.

Over nine years, beginning with the failure at Makara in 1970, I had laid the basis for the Baring Head atmospheric CO_2 programme;

it was up and running, internationally recognised, and I had published the first part of the record. I'd poured everything I had into the work and was proud of what I had achieved, yet I felt more than a little wistful leaving the programme to others. In November 1978, with the handover procedures complete and our goodbyes said, Irena and I left New Zealand for Germany, travelling via Scripps.

When I met up with Keeling he was beaming. 'Congratulations David,' he said, and shook my hand. 'Great news on the scholarship. I'll come and visit you in Germany when I can. I know Dieter Ehhalt well – he is a brilliant atmospheric chemist, one of the best in the world.'

After everything we'd been through with the trials of Baring Head, here was my hero congratulating me in person – it was a proud moment.

Too soon it was time to leave, once again with Peter and Chris waving us goodbye. But this time was different: we were flying even further from home, to face an unknown future in Jülich, West Germany.

CHAPTER 10

JÜLICH, WEST GERMANY

'What's that terrible smell?' asked Irena. Exhausted from our flight, we'd fallen asleep immediately on arrival only to be woken at 5am by dozens of tractor engines revving in the street outside the hotel.

'It's like beetroot being boiled,' I replied. 'It's horrible!'

My guess was close – the smell came from an enormous sugar beet factory on the outskirts of Jülich, less than a kilometre from the hotel. We'd arrived towards the end of the sugar beet harvest and in the street outside our hotel were dozens of tractors towing trailers piled high with sugar beets harvested from the surrounding fields. At the peak of the harvest there were long processing delays, and tractors and trailers jammed the streets from early in the morning until it got dark at about 4pm.

Unable to sleep, we got up, had coffee, and spent the morning walking the streets of Jülich. The pungent smell permeated every part of the small town, caught in our throats and stuck in our clothes. Bone-chilling winter air gripped our bodies as we walked past grey ugly buildings, built in a post-World War Two architectural style.

'Have you noticed – there's no wind?' said Irena. 'I never thought I'd miss the Wellington wind!'

In World War Two, Jülich was obliterated by allied bombing, and the majority of the houses and apartments within the town were rebuilt during hard economic times after the war. The population had to make do, often recycling materials scavenged from the rubble. There was no thought of architectural flair. But, by the time we arrived, Germany had become one of the wealthiest countries in

Europe and many attractive houses were being built on the outskirts of the town. However, we spent the morning in the town centre, wandering past grey buildings, grey trees and grey stone walls under a grey sky. And it was bitterly cold. The blue clear skies we had left behind were replaced by low hanging cloud. Everywhere, there was a layer of aerosols and the air felt dirty – this was the edge of the Ruhr industrial area, a region ringed by giant brown coal power plants with huge stacks belching particulates, gases and steam into the air. The gloomy skies were to last another six weeks before we caught a glimpse of what passed for the sun, a dull red orb barely rising above the southern horizon.

We stumbled back to our hotel room utterly dejected. Irena threw herself onto the bed and burst into tears. 'What have we done,' she cried. 'Leaving New Zealand, for this!'

That afternoon, we were gloomily drinking tea in the hotel lobby when a tall, casually dressed man approached us.

'Dave and Irena Lowe?' he asked. The man introduced himself as Ulrich Schmidt from the atmospheric chemistry institute at the Kernforschungsanlage. 'That means "Nuclear Research Centre",' he added, 'but we just call it the KFA, and the atmospheric chemistry institute has very little to do with their research.'

Ulrich offered to take us to meet Professor Dieter Ehhalt. He drove us into the country and through a forest to the KFA, a nuclear research facility with 3500 employees, ringed by razor wire and patrolled by armed guards with dogs. Getting through the main gate involved a tight security check, then Ulrich continued through to the atmospheric chemistry institute.

There we met a smiling Dieter Ehhalt, wearing jeans and with his feet up on a desk. Several people in New Zealand had warned me about the formality of the West German higher education system. One even said I would have to wear a suit every day. I didn't own a suit; I'd only worn hired ones for graduation ceremonies. Yet here was this laidback professor looking ready for a neighbourhood barbecue! What a relief.

'Welcome to Jülich and the KFA,' he said. 'I know you've come an incredibly long way.'

Dieter explained there were a few administrative formalities to go through, which Ulrich would help us with. I was a little puzzled, but in less than a week I discovered exactly what they'd been alluding to – I soon learned that administration is a way of life in Germany and you adapt to the formalities or drown in a bureaucratic soup.

After the introductions Ulrich took us to pick up our suitcases and move into the apartment he'd booked for us at the KFA Gästehaus. At ten floors, this was the tallest building in Jülich. Our apartment was on the eighth floor, with a commanding view of the flat countryside around Jülich.

'Give me a call if you need anything,' said the ever-smiling Ulrich. We thanked him for his help as he left. The apartment door closed behind him and we were alone. For the first time since arriving in Jülich, we laughed with relief. Finally, after a warm welcome it felt like everything was going to work out. And so our three and a half years of new friendship and life-changing adventures in West Germany began.

From my first day in the lab at the Institut für atmosphärische Chemie 3, or simply ICH3, which is what everyone called the KFA's atmospheric chemistry institute, Ulrich was a constant mentor and companion. Although Dieter Ehhalt was my chief supervisor and registered as my professor at the University of Cologne, it was Ulrich who advised me on a day to day basis, picked me up during countless setbacks, insisted that I spoke German and was always there for me.

In 1966 Ulrich had been one of the first students in the world to study atmospheric chemistry, studying as a master's candidate with Professor Christian Junge at the University of Mainz. In 1968 Professor Junge was awarded the inaugural position as director of a new air chemistry division at the prestigious Max Planck Institute for Chemistry in Mainz. Junge offered Ulrich the chance to do a PhD with him, and under Junge's supervision Ulrich went on to develop the first technique for measuring minute levels of hydrogen in the atmosphere. His completed thesis and 1975 paper on molecular hydrogen in the atmosphere were widely cited, earning him accolades and worldwide attention. In 1976 he joined Dieter Ehhalt at the KFA, becoming his deputy in 1978.

When I discovered that not only did Ulrich know 'the father of atmospheric chemistry' but had actually *studied* under him, I was really encouraged. Despite our initial gloomy first impressions of Jülich, I knew that I had made the right decision to study atmospheric chemistry at ICH3. What remarkable backgrounds both Dieter Ehhalt and Ulrich had – yes, I already had a good basis in atmospheric science, but I felt humble as I began my PhD under the supervision of these two world experts.

And they had certainly assigned me a challenging PhD topic. While atmospheric CO_2 is a powerful greenhouse gas, it is not the only one. Methane is second to CO_2 in its effect on global heating. Like CO_2, it is increasing due to emissions from a variety of industrial and agricultural sources. There are grave concerns that, as well as its direct impact on atmospheric chemistry, positive feedback caused by global heating is leading to the release of even more methane from permafrost, ocean sediments and other sources.

In 1974 Dieter Ehhalt published a farsighted review paper examining what little was known at that time about the sources and sinks of atmospheric methane. The behaviour of methane in the atmosphere is very different to that of CO_2. Whereas excess CO_2 produced by fossil fuel combustion has a long lifetime in the atmosphere, hundreds to thousands of years, methane has a relatively short lifetime, about ten years, before it is destroyed in the atmosphere by a sequential series of chemical reactions.

It was an intermediate part of this series of reactions that Dieter Ehhalt wanted me to investigate for my PhD. Methane is removed from the atmosphere by incredibly small quantities of OH, or hydroxyl, which is a free radical produced by the action of sunlight on water vapour. A free radical is a chemical species containing an unpaired electron, making it highly reactive and unstable. Free radicals are responsible for many of the important reactions in atmospheric chemistry. When atmospheric methane is 'attacked' by OH, a series of chain reactions eventually leads to small amounts of CO_2 being produced. At the time of my thesis work, the reactions had been modelled theoretically but, due to their speed and the tiny quantities of the reactants involved, experimental studies were virtually non-existent. Dieter hypothesised that it should be possible to gain insight into this atmospheric methane removal process by measuring the longest-lived intermediate species in the series of reactions. This was atmospheric formaldehyde.

When he told me about his idea, I was immediately interested. Here was a chance to make a contribution to understanding the behaviour of another very important greenhouse gas, methane.

He, Ulrich and I sat in his office on the first day of my PhD programme. 'I assume the formaldehyde needs to be measured in clean air?' I asked.

'Yes, that's right,' he said. 'Measuring it in polluted air is interesting but does not help us understand the process.'

'How much formaldehyde do you think there is in the clean atmosphere?' I asked.

'We're not really sure, but it might be a maximum of about 0.2 parts per billion,' he said with a smile, anticipating my reaction. 'And sorry, it gets even harder – we think its lifetime is only a few hours in the sunlit atmosphere because it's being both generated and removed by photochemistry. But that's also uncertain.'

'Has anyone been able to measure it?' I was beginning to realise just how challenging my PhD would be.

'Well, there are quite a few measurements in polluted air,' said Dieter. 'But nothing in clean background air. If you can figure out how to measure it in remote areas, that will really help us to understand what's going on.'

Ulrich noticed my despondent look. 'We have a few ideas. How about we go down to the lab and I'll talk them through with you.'

I stumbled out of Dieter's office in disbelief. At only 0.2 parts per billion (ppb), atmospheric formaldehyde had a concentration more than a million times lower than CO_2, not to mention it only survived in the atmosphere for a few hours at a time. A challenge indeed.

Our first few weeks living in Jülich were extremely unpleasant. The temperature remained bitterly cold but without snow, and walking outside was often dangerous due to what the Germans call *Glatteis*, a mirror-smooth, almost invisible layer of ice that forms on hard surfaces like roads and footpaths. The local hospital was full of people with broken noses, fingers and hands who had slipped and fallen on the ice. Inching our way round, one short slithered footstep at a time, became the only way we could be sure of not slipping and injuring ourselves. Fortunately the main supermarket, Kaisers, was only about 200 metres from the Gästehaus. Getting to the atmospheric chemistry institute at the KFA was easy for me; I used a local bus which went virtually door to door. When the bus arrived at the stop outside the Gästehaus, it would slide on the icy street for several metres towards the waiting passengers, with all four wheels locked. Sometimes the drivers would miscalculate and we'd have to jump out of the way as the bus careened onto the footpath where we'd been standing seconds before.

Each day seemed to be a matter of survival under grey skies as we gradually learned to use the supermarket, post office, pharmacy and other shops. The Gästehaus had about forty apartments and we started meeting some of the other residents from around the world – Japan, India, Iran, Iraq, China, England, South Africa, Germany, the US and Australia. Many had Jülich 'survival stories' and sometimes misleading advice on how to get by.

We soon learned to be sceptical of a lot of what we were told. 'Did you know that Kaisers sells beer in crates of twenty half-litre bottles?' an excited Australian told me not long after we had arrived. 'You pay a five Deutschmark deposit on the crate and bottles, and they give you the money back when you return them,' he continued. 'The beer is bloody good and it's quite cheap!'

German beer has been brewed for literally hundreds of years following a strict *Reinheitsgebot*, a purity law which mandates the ingredients in beer to just four: yeast, hops, malt and water. In New Zealand, the US and many other countries in the 1970s, commercial beer was a random chemistry set with proprietary amounts of foaming agents, foam stabilisers, colouring matter, preservatives and other secret chemical additives tossed into the brew.

After getting the recommendation I lost no time going down to Kaisers and buying a crate of beer. It was absolutely delicious, but it was warm. Back then there were no walk-in beer fridges in liquor outlets, and in our apartment, we only had a tiny fridge with no room for more than one or two bottles of beer. Remembering how cold it was outside, I suggested to Irena we use the balcony.

That night the temperature dropped dramatically from the daytime maximum of about -2°C to -18°C. The beer froze solid and blew all the caps off the bottles. When we checked the following day, we were greeted by frozen trails of beer foam and almost empty bottles. For a student, this was a tragedy indeed. At least, we thought, we could get our deposit back from the supermarket.

'Good afternoon. Could I have the five Deutschmark deposit back on these empty bottles and crate please?' I asked, smiling at the cashier in Kaisers. I was proud of the German sentence I'd practised.

The cashier proceeded to reply in a completely incomprehensible sentence. I was dumbfounded – the only two German words I could understand were 'vegetable department', a link which eluded me. I repeated my question. The receptionist shouted at me, louder and faster this time, and still I had no idea what to do. By this time a queue was forming behind me and I could hear muttering that included the words *Doofer Sack* – literally, 'stupid sack'. I had no choice but to walk out, leaving the crate of empty bottles behind.

'Not sure what's going on there,' said the Australian at the Gästehaus when I told him what had happened. 'To be honest, we haven't tried it yet. So far, we've just piled our empty crates and bottles up at the back of our apartment.'

It was to be another two weeks before Ulrich solved the mystery for me. 'You take your crate to the vegetable department, where they give you a receipt for the deposit. You take the receipt to the cashier and claim your money back,' he explained, laughing.

I admitted the system made some sense. 'But I could only understand the words "vegetable department". Everything else sounded like *Wot is dat dan duh duh . . .*'

'That's because it does sound like that,' he said. 'What you heard was a mixture of local dialect and High German. I don't understand it either, and that goes for most of the people at the KFA who have come from other parts of Germany and overseas.'

'Pity they had to shout at me,' I said.

'Did you know that, if I go to England and ask a question in English, and they reply with something I don't understand, they'll sometimes just continue to repeat what they just said louder and faster, often ending by shouting. It happens both ways.'

Ulrich was absolutely right. And New Zealanders tended to respond to foreigners in exactly the same way that the cashier at Kaisers had to me. Of course, this revelation didn't get round the problem of understanding the Jülicher Platt dialect, a problem shared by everyone from outside the district. However, Ulrich's advice was that, if I persevered with High German, the response would eventually come back in that, albeit with a strong Jülich accent.

There are a lot of very expressive words in German, and a relevant one is *mitmachen*. This can be loosely translated as immersing yourself in the things and activities around you. Once we appreciated this, our lives in Jülich became more enjoyable. The closed enclave of foreigners we had met at the Gästehaus had been a useful and comfortable beginning. But as we became more confident with our spoken German and learned about Jülich, we began to drift away from the Gästehaus, preferring to live and mingle with the many German families associated with the KFA who lived elsewhere in the town.

One night, after a day in the lab at the atmospheric chemistry institute, it seemed a little warmer outside – about 5°C, positively tropical compared to the sub-zero temperatures of the previous weeks. Irena and I dressed up in warm coats and hats, sat out on the balcony and enjoyed a couple of beers that I had just bought from Kaisers. We grinned at the knowledge that I would be able to return the crate of empties via the vegetable department without having the cashier shout *Wot is dat dan!* at me.

My early days at the atmospheric chemistry institute were spent searching past literature for effective techniques to measure formaldehyde concentrations. Hour upon hour passed reading abstracts of scientific journal articles stored on microfilm, which could be magnified using a special reader. Finding what I was looking for was a laborious process, tiring and often frustrating. I'd already done an initial literature search on potential techniques for formaldehyde before I left New Zealand, and Ulrich had also done some preliminary searches. Between us we had discovered that in the late 1970s, there were only two sets of measurements reported for formaldehyde in clean air. One of these was a measurement series that had just been completed by Uli Platt, a brilliant post-doctoral fellow working for Dieter at the institute. He'd used a laser technique which obtained an average concentration for the formaldehyde contained in air in a 5-kilometre path-length above a nearby field. The method was very sensitive, but my PhD topic required measurements of formaldehyde in *vertical* profiles of clean background air as well as in mid-ocean air. Samples would have to be obtained from aircraft and ships in remote locations. The second published technique was based on a 'wet chemical' extraction procedure developed during World War Two for determining the formaldehyde levels inside submerged U-boats. This wartime study seemed thorough, but replicating the technique was impossible.

Ulrich and I had also found other techniques designed to measure much higher levels of formaldehyde such as what's produced in forest fires and found in tobacco smoke. Whether these techniques could be adapted to measure the much lower concentrations expected in clean background air seemed doubtful. But neither of us had any other options, so I began experimenting with basic wet chemical techniques.

We focused much of our attention on a relatively new technique called high-performance liquid chromatography (HPLC), which at the time was most well known for detecting performance-enhancing drug levels in athletes' blood and urine tests at sporting events like the Olympics. The basic principle was the same: we would need to find a very sensitive marker for formaldehyde which would make the molecules visible to the HPLC. We soon discovered a promising marker, an organic compound we called DNPHzine. But my excitement was premature, as I spent many wasted weeks battling problems with instability in the HPLC, poor sensitivity and contamination in the organic solutions I was using. Every day my measurements would give a different result – the technique seemed to be completely unreliable.

By the end of 1978, despite experimenting with a number of methods and throwing everything I had at the project, I'd made zero progress. Compounded by the grey gloom and bitter cold of our first winter in Jülich, I was frustrated and depressed. PhD projects, by their very nature, involve tackling topics which have not been investigated before. Their final outcome is almost always unknown. Even the best supervisors in the world propose PhD programmes that are doomed to failure – was mine to be one of them?

CHAPTER 11

A TALE OF SERENDIPITY

My search continued for a technique to determine low levels of atmospheric formaldehyde. The handwritten lab books I used during those first few weeks record day after day of frustrating failure, and it became clear that the HPLC technique was never going to show the sensitivity we needed.

One night on an HPLC training course in Darmstadt, a city about 300 kilometres from Jülich, I sat in my hotel room in familiar despair. I thought back to the years I'd spent overcoming other obstacles. For no apparent reason my mind wandered back to an early student experiment where we had increased the efficiency of a simple alcohol distillation using some strange glass rings; were they called Raschig rings?

Suddenly I had an idea – the efficiency increase must have been due to the very large surface area on the rings. I grabbed a piece of paper and quickly calculated that even a small volume of rings could provide a large surface area for sampling low concentrations of a gas like formaldehyde. Inspiration comes calling at the strangest of times. By taking a step back from the techniques I'd been pursuing and using something like those weird rings to trap the formaldehyde from huge air samples directly into a sampling solution, I realised I might be able to avoid the terrible wet chemical technique. I stayed up late making calculations and checking the chances of contamination or random errors. That night, I couldn't sleep. I got up a couple of times in the early hours and redid my calculations; the results were the same. Everything looked really promising and I couldn't wait to get back to the lab.

When I returned to Jülich and told Ulrich about my idea, he was immediately interested. Thankfully, and perhaps as no surprise given the chemical engineer who'd invented Raschig rings was German, the stores at the KFA had a large selection of them. I lost no time designing an air sampler based on upright glass cylinders filled with the rings bathed in a small volume of DNPH sampling solution. The glass blower at the KFA had already built similar vessels and he had no problem making one from a sketch I'd given him. As soon as he'd completed it, I set it up on a balcony at the back of the lab and, improvising, used a car vacuum cleaner to suck a whole cubic metre of air through the sampling solution in the Raschig rings.

Trembling, I extracted a syringe full of the solution and injected it directly into the HPLC.

'Yes!' I yelled. Ulrich came running into the lab.

'Look,' I said, pointing at the chart. 'It's an enormous formaldehyde peak.'

'You got that from outside air with the new sampler you just designed?' he asked. 'This could be the breakthrough we've been looking for!'

When Dieter Ehhalt came into the lab that week, Ulrich and I were still getting to grips with the new technique. He was clearly very pleased, and congratulated us with a slap on the back. We would have to do a whole lot of control tests, but the technique could be ready to begin sampling in a couple of months. Dieter congratulated us again, and left.

Ulrich looked flabbergasted. It took him a couple of minutes to compose himself before he said, 'You know, Dave – that's the first time he has ever slapped me on the back and called me by my first name. He's obviously very excited.'

The more German I spoke and understood, the more I enjoyed life in Jülich and the easier things became. I was beginning to be able to read newspapers and understand the news on TV, and Germans felt comfortable talking to me and asking me my opinion on a wide variety of topics. For a non-language student like me this was a revelation. It opened my mind to concepts that were not available in English; it was like being in a room where suddenly new windows opened up on scenes that you'd never seen or been able to appreciate before.

Ulrich and I had an agreement that, for most of the day in the lab we would speak German, but when I got tired later in the afternoons we would switch to English. It was a good arrangement for both of us – my German was improving and Ulrich was keen that I corrected his English. But I knew that to improve my grasp of the language, preparing for and taking a German proficiency exam was the way forward. It was also – more importantly – a requirement in order to matriculate for my PhD.

The University of Cologne's proficiency exams for foreign students were held several times a year and preceded by an optional six-week preparation course held three days a week in the languages department. Ulrich and Dieter agreed I should attend the first available course and then sit the proficiency exam.

Getting to the university from Jülich by public transport was a five-hour round trip which began at 5:30am with me running through the snow in the dark to the railway station, catching two trains and two different trams, and finally arriving at about 8am for the first lecture of the day.

On the first morning of the German language preparation course I joined about 300 foreign students in a lecture theatre. We had a short welcome by a senior lecturer, and that's where my understanding of what was going on ended. The lecturer continued with an elaborate and, to me, virtually incomprehensible speech for another five minutes, at which point she stopped abruptly and waited. To my surprise more than half of the students in the lecture theatre got up and left. I stayed, understanding almost nothing for the remainder of the day but picking up a large amount of written material including grammar and comprehension exercises, which I went through carefully with Ulrich and others at the atmospheric chemistry institute the next day.

It wasn't until the following week that another student at the language course told me what had happened on our first day: the lecturer had said that unless you'd been in Germany for at least six months and preferably had spent a year just learning German, then this course would be too hard and you should leave. I hadn't understood even that so had stayed – dumb Kiwi!

I persevered for the next month, staying a week at a time with German friends who lived in Cologne to save the hassle of public transport from Jülich each day. My friends were determined I was going to pass the exam and rigorously tutored me, going through the written and oral exercises from the university each day. It was gruelling full immersion in the language but I was gradually getting up to speed. Ulrich was also very helpful but occasionally got angry with the written material from the university.

'What brain-burned idiot designed this crap?' he said abruptly one day when I showed him the latest exercises I had to do. 'No one in Germany speaks like this – it's far too academic. They should be teaching you stuff that's useful for everyday life.'

After six weeks, exhausted from the travel back and forth and my head full of German phrases like 'Due to an epidemic of sore throats, more and more officials were absent from work', it was time to sit the proficiency examination. Perversely, I remember quite enjoying the written exam; the comprehension piece was about a female gorilla who could communicate in sign language. The exam question went on to ponder questions of ethics, integrity and morality in gorillas.

Later that week I was told that I'd got about 70% in the written exam, meaning that in order to pass I would have to take an oral exam with one of the lecturers. Typically, these sessions lasted for up to thirty minutes for each student – the rumour was that the longer the oral exam took, the more likely it was you were going to fail. When I arrived for my exam, the orals were running late and I saw three students before me come out of the exam room in tears – they'd failed.

When it was finally my turn to be tested, I was nervous – to put it mildly. There were two examiners, and we began a conversation which first involved questions about the gorilla question from the written exam. One of the examiners was actually a physical scientist, and asked me about the atmosphere and the atmospheric chemistry institute. I relaxed and started to enjoy what was turning into a lengthy conversation, eventually noticing that we'd been going for about twenty-five minutes. Bugger – it had lasted too long. Was this a failure? But the examiners read my mind. They told me not to worry, they'd enjoyed our conversation and determined my command of the language was more than enough to begin working on my PhD. I had passed.

My elation over the breakthrough with the atmospheric formaldehyde technique was shortlived. Yes, I'd shown that the method could work, but I soon discovered the results were unreliable. Measuring formaldehyde in the same air sample, the next day or even just a few hours after collecting it, produced different results. I was bitterly disappointed.

My handwritten lab books from those years record month after month of fruitless experiments. As if these trials were not enough, the HPLC started playing up: fuses blew, the electronic integrator stopped working, the motor driving the chart paper jammed and the ultraviolet detector blew up – all in the space of a few weeks. I'd battled with years of equipment problems at Baring Head; at least in Jülich I didn't have to contend with gale-force winds. Somehow, I just kept going. Giving up was never an option.

You know the phrase 'There must be something in the water'? To make the sampling solutions for my technique, I had been using double-distilled water to remove any last traces of formaldehyde. This was critical. The last thing I wanted was to swamp the atmospheric samples with background formaldehyde in the sampling solution. I'd calculated that using double-distilled water would reduce the levels close to zero. I'd also tested the water, or so I thought. One morning the water distillation plant in the lab failed and, to continue with my experiments, I was forced to use water from a pure water plant based at another chemistry institute in the KFA. I was amazed to find that the first runs with the water from the other plant produced much lower levels of formaldehyde.

In hindsight it was such a simple thing – the water I'd been using, even though it had been double-distilled, still had relatively high background levels of formaldehyde. Somehow my early tests had failed to identify this. Worse still, the amount of formaldehyde changed with time, accounting for some of the variability I'd seen. Ulrich lost no time in ordering a state-of-the-art water purification system for the lab, which led to a dramatic improvement in my results. It was one of the first of many problems I had to overcome, but it strengthened my will to keep going.

By the end of April, spring was with us. Irena and I spent a lot of time riding around on secondhand bikes we had bought from neighbours. Jülich is surrounded by sugar beet fields with small isolated pockets of forest. Paths and lanes allow farmers and forest workers to access their land, and give walkers and cyclists a safe alternative to narrow German country roads. We explored the entire region on our bikes, often taking a picnic lunch with us. Riding through the forests in spring was a revelation – the brilliant green of fresh new leaves astonished us after a long winter. We'd spend hours out amongst the new growth on chestnut, beech and oak trees.

As well as the bikes, we bought a secondhand bright yellow Audi 80, and used it for day trips. With the car we soon began to make forays into Belgium and the Netherlands, usually taking our bikes with us. Our knowledge of this part of Europe was exploding; the more we experienced, the more fascinated and enthusiastic we became with the environment around us. At the same time, we were meeting more and more Germans, who introduced us to places further away – the Eifel hills, the Mosel and Rhine river valleys, the cities of Cologne, Aachen and Düsseldorf and even further away, Heidelberg and Stuttgart in the south. With the help of our friends we moved out of the Gästehaus into a small two-bedroom apartment on the ground floor of a block on the outskirts of Jülich.

By summer 1979 we were starting to be inundated by visitors, mostly from New Zealand. Almost by default, we became an enclave where young Kiwis on their OEs could hang out with two New Zealanders who spoke the language and could drive them places. A lot of the visitors we knew beforehand, but some we had never heard of. I received one memorable phone call from Trevor, who introduced himself as 'a friend of a bloke' who met my brother at a party in Auckland and asked if I could pick him up from the airport and put him up for a night. We had almost thirty visitors that year, and while we thought being a base for travelling Kiwis was fun, it was a constant source of amusement to our German friends.

A very special visitor was my mother, accompanied by Irena's youngest brother, then only twelve years old. Mum had never been to Europe before and absolutely loved what she saw in Germany. We took her by train to Paris to see the Eiffel Tower, a sight she described as one of the highlights of her life.

Another of our visitors was Irena's brother Jan, who was living and working in London as an accountant. On his first visit in June 1979, we decided that the three of us would drive to Poland. At that time Poland was still under communist rule, heavily regulated by the Soviet Union and a journey there was not without danger. And to get to Poland by car, we would have to cross the border and drive 400 kilometres through heavily policed East Germany. None of my colleagues at the institute had attempted a journey like this. As usual they joked, *Doofer Sack!* Ulrich was a bit more forthcoming.

'Be very careful and obey all the road rules when you cross the border,' he said. 'And whatever you do, don't take any Western magazines or journals with you. You can take a car manual and a road map of Poland, but make sure that it only has Polish names for the cities.'

We bought a road map of Poland at a bookshop with the place-name index in two parts: one in Polish and the other in German, a section that could be removed. A few days later the three of us arrived in the yellow Audi at Bad Hersfeld, one of the road border crossings into East Germany. After about two hours in a queue, a heavily armed guard approached.

'Do you have any machine guns, rifles, pistols or weapons with you?' he asked.

'Er, no – nothing at all,' I said, a bit intimidated.

'Drive over to that shed, then I want all three of you out of the car! Wait beside the shed until I tell you to move!' he said, indicating with his rifle.

We watched as he took apart all of our carefully packed luggage. He turned Jan's backpack upside down, tipping the contents over the ground. After about five minutes he returned looking very angry.

'What's this?' He held up a small book. I realised in dismay it was an Agatha Christie mystery called *Elephants Can Remember* which we'd borrowed from the Gästehaus library.

The guard took the book to the main guard post. An hour later he returned, and told us we could go. He almost smiled. 'I've just talked to the main command in Berlin. They investigated and told me that this book is actually rubbish!'

I threw the book into the car and with relief we drove off, across East Germany to the Polish border at Gorlitz. Of course, the inevitable happened – we forgot about the novel lying on the back seat. After an explanation to yet another guard, this time we were released relatively quickly and drove into Poland without any more incidents. I lost no time tossing *Elephants Can Remember* into the nearest trash can – the book never made it back to the Gästehaus library.

We spent an astonishing two weeks in Poland with Irena and Jan's relatives. What an eye-opener Polish communist rule was to the three of us. All of our values were challenged – Irena's relatives seemed incredibly poor, yet the state paid for hospital and dental care, childcare and three weeks' annual holiday at a kind of health spa. The state owned their apartment and almost no one owned a car. The shops were all state controlled and so never went out of business or held a sale. A lot of stock simply deteriorated to the point it had to be thrown out. We returned to West Germany staggered by what we had seen but wiser from the comparisons and contrasts we could make between these two countries.

Much of the research at the institute involved developing techniques to understand the behaviour of the OH free radical. This key atmospheric constituent, present in vanishingly small quantities in the atmosphere, is a major driving force in atmospheric chemistry. My research into atmospheric formaldehyde would contribute to this knowledge; methane, a major greenhouse gas, is removed from the atmosphere by the OH free radical in a chain of photochemical reactions that form formaldehyde, among other trace species. The OH free radical cleanses many pollutants from the atmosphere and is often known as 'the atmospheric detergent'. If the OH radical were not present, then the lower part of the atmosphere would rapidly fill with smog and other pollutants. This includes carbon monoxide, which would reach toxic levels.

What I find truly amazing is that OH achieves this at minuscule atmospheric concentrations. For perspective, the total area of the United States is about 10 million square kilometres. In that huge area, the amount of OH typically present in the atmosphere would be represented by a single paving stone of about one square metre. Describing the OH radical as a needle in a haystack is an understatement.

One research group within the atmospheric chemistry institute measured very low levels of atmospheric trace gases by beaming lasers and other powerful light sources through air over long path-lengths into different detectors. One of their research goals was to detect the OH radical in the atmosphere. This group, which included a post-doctoral fellow named Uli Platt, went on to perfect a technique known as differential optical absorption spectroscopy. This apparatus is now used worldwide in a variety of applications, including studies of volcanic gases and air pollution. The group's research was groundbreaking and they were recognised as world leaders.

Meeting Uli Platt from the laser group led to a lifelong friendship. There was an immediate spark and recognition between us. We could joke and laugh as we learned from each other. To him, at first, I was the Kiwi *Doofer Sack*, and to me he was a heavily bearded, long-haired guy with thick glasses, a deep intellect and a wicked sense of humour. He could do (and still does) blindingly fast mathematical calculations in his head. I could do the same and it was a lot of fun challenging him. But I was not in the same league – he was much faster and made far fewer mistakes.

One day Uli and I were in the lab drinking truly terrible coffee and he asked how my project was going.

'It's this bloody wet chemistry!' I complained. One of my biggest problems, I said, was that to improve the signal-to-noise ratio for low formaldehyde concentrations, the only way through would be to use huge air samples – perhaps as big as two cubic metres – and at high flow rates. I groaned. 'How the hell am I going to get all the formaldehyde in an air sample into a sampling solution?'

'You said that you improved the efficiency of your technique using Raschig rings?'

'Yes,' I said. 'The rings were a big improvement, but there are still stability and efficiency problems.'

'You know,' he said, 'It was a completely different application but I remember that my supervisor in Heidelberg used Raschig rings in a rotating glass drum to collect samples for radiocarbon in atmospheric CO_2 – could be worth a look?'

It was another eureka moment. I modified the design of my sampler and gave the plans to the glass blower. He took one look, laughed and told me that he thought he had seen everything but this was really weird. He said he loved a challenge, and would build it.

The new sampler required some innovative ways of rotating the glass drum using a small motor and gearbox, as well as gas-tight connections allowing air to be sampled through the drum while it was turning. I was used to improvising and building new equipment, and with Ulrich's help tracked down suppliers for the various components I needed. I became a regular customer at a car parts store in Jülich, where I bought car vacuum cleaners. Before the design was complete I'd burned out quite a few, overheating them with the large volumes of air I was using. The sales staff at the store got to know me. 'You're the best customer we've ever had for car vacuum cleaners!'

The first runs of the new rotating Raschig ring sampler were a big improvement, with a large atmospheric formaldehyde signal providing a lower detection limit of 0.01 ppb, a factor of 10 better than I needed for the background air measurements that Dieter wanted. The new rotating sampler also made it easier to identify remaining problems. For example, was the sampler collecting all the formaldehyde in an air sample? I needed to do a lot more control experiments, and for these I designed a complicated double rotating drum system. When I took the plans down to the glass blower, he threw his hands in the air. Laughing and shaking his head, he admitted this design was beyond his abilities and directed me to the main KFA glass workshop, about a kilometre away on the other side of the campus. When I arrived with my plan, I was amazed to see six glass blowers in overalls sitting at a table, drinking beer and eating plates full of sausages and sauerkraut. It was 11am.

'Excuse me,' I said. 'I was wondering whether one of you might be able to make the glass sampler I've drawn on this plan?' There was a chilly silence.

'Can't you see we're in the middle of our second breakfast!' shouted one. 'Get out!'

I trudged back to the atmospheric chemistry institute. When I told Ulrich what had happened, he thought it was hilarious.

'The second breakfast is a very important custom for many workers in Germany,' he said. 'But I'm not sure about drinking beer in the morning. I'd go back in the afternoon.'

When I went back to the main glass workshop just after 2pm, all six glass blowers were working. One of them took my design and said, 'Complicated, but okay. Come back in three days.'

*

For the next two months I worked every day with the new sampler, gradually refining the technique to the point where the results were stable and reproducible, and I could trust the system on a field campaign. At room temperature the samples were stable for a few days, but in a fridge they were stable for months.

Dieter was excited by the refined technique and wanted me to get clean background air data as soon as possible. He was convinced that, made in the right location, my formaldehyde measurements would contribute to determining the lifetime of methane in the atmosphere, data essential to understanding its role in increasing global heating.

Both Dieter and Ulrich were concerned about the changing chemistry of the atmosphere. They were fully aware of the disastrous implications of increasing atmospheric CO_2. But they were also worried about other human impacts on the atmosphere, particularly those that could affect the concentration of OH in the lower atmosphere and ozone levels high in the stratosphere. This was before the discovery of the 'ozone hole' above Antarctica, but new research was already suggesting that the widespread use of industrial chemicals like CFCs – used as propellants in aerosols and as refrigerants – could be a serious problem, causing a chemical breakdown that would deplete stratospheric ozone. One of the researchers on this, Sherwood Rowland, was invited by Dieter to lecture at the atmospheric chemistry institute. I remember his presentation sent a chill through me – here was yet another way that humans could affect the atmosphere on a planetary scale. In 1995, Rowland and researchers Mario Molina and Paul Crutzen went on to win the Nobel Prize in Chemistry for their findings. A few years earlier their research had helped establish the Montreal Protocol, an international agreement banning the production of CFCs and other stratospheric ozone-depleting chemicals. The agreement has been successful, and the annual Antarctic ozone hole is beginning to shrink, averting a potential disaster for the human race. If only early efforts to reduce carbon dioxide and methane emissions could have been so successful.

Ulrich suggested building my sampling gear into a small twin engine research aircraft, a Dornier Do 28 run by the German aerospace organisation, DLR. The aircraft was based at a small airport not far from Munich, and three of us made the 600-kilometre trip in a KFA van with my equipment. It took us two days to mount glass air intakes on top of the aircraft's fuselage and couple these to the battery-powered rotating drum sampler in the aircraft's cabin. I'd calculated that the pressure created by the moving aircraft would produce a flow rate of 600 litres per minute, of which only 40 litres per minute would be taken by my equipment. The idea was to avoid any contamination problems with formaldehyde decay on the glass surfaces of the air intake. But I was worried I'd made mistakes with the fluid dynamics equations I'd used for the calculations.

'No problem,' said the ever-practical Ulrich. 'We'll test the intakes with my car.'

Ulrich drove me and the sampling tube out to the nearest Autobahn, and I checked the flow rate with the tube pointing forward out of the front passenger window while he drove at ever increasing speeds. There were no speed limits on these motorways, and we tested the intake while driving at 170 kilometres per hour. At this speed the flow rate was already well over 300 litres per minute and was obviously going to be fine at the 250 kilometres per hour cruising speed of the aircraft. 'The things we do for science!' said Ulrich afterwards, a grin plastered ear to ear.

After we'd installed the equipment, we met the pilot and the aircraft at Düsseldorf Airport. The flights were to be in the troposphere, which is the part of the atmosphere that supports all life and begins at ground level. Depending where in the world you are, the troposphere can reach a height of at least 10 kilometres above the ground. At this point, an abrupt increase in temperature forms the boundary between it and the stratosphere. In the troposphere, regions where the temperature suddenly increases with height can also cause localised inversion layers which trap noxious gases close to the ground, leading to dangerous smog above many cities. In rural areas, smoke from fires is often trapped by inversions close to the ground, appearing as well-defined smoky layers above villages in valleys. Dieter wanted me to fly above any local inversion layers to be sure that I was sampling formaldehyde in air uncontaminated by pollution. He suggested that I use temperature readings from the aircraft to find the height of any inversion layer and sample both below and above it.

On four separate days I flew up to altitudes of 5000 metres above the Eifel hills. Each air sample took an hour to process, and I managed to collect a total of fourteen over the four days. The views of the Eifel hills were breathtaking. Brilliant autumn colours spread out beneath me in the forests covering the hills below, with blue lakes threading along valleys. It was an experience I've never forgotten. After I'd completed air sampling on each flight, the pilot would invite me to sit in the co-pilot's seat for the return to the airport. At Düsseldorf the landings were thrilling. Once we got close to the airport, the pilot would point the aircraft steeply down towards the runway and increase the engine speed, cutting the throttle just as the plane touched down.

It was almost twelve months to the day since we'd arrived in Jülich and the hard grind had paid off. The sampling equipment performed flawlessly, producing the first ever measured vertical profiles of formaldehyde in the troposphere. As predicted, the temperature readings showed a strong inversion at about 3000 metres. Samples taken above this had very low formaldehyde concentrations. Simultaneous CO_2 and methane measurements showed that the samples had been collected in clean air. Dieter congratulated me and told me that I should publish the data and technique immediately. Over the next couple of months, I worked on an article entitled 'A New Technique for Measuring Tropospheric Formaldehyde' with Dieter and Ulrich, which we published in the well-regarded American Geophysical Union scientific journal, *Geophysical Research Letters*.[10]

Earlier in the year Irena and I had decided to start a family, and so we were thrilled when a doctor told us Irena was pregnant and expecting our first child in December. We looked forward to the birth with mixed feelings. We knew our lives were going to change forever. The due date was mid-December. My birthday is on 16 December, so we joked about combined father and child birthdays. Getting to grips with the German healthcare system was challenging, but we had a lot of help from our German friends. We attended German prenatal classes and used to joke afterwards that we knew all of the German words to do with having babies but not the New Zealand ones.

As the days lengthened and turned to summer, we spent most weekends exploring the local area on our bikes or further away by car. Nearby, the heavily forested Eifel hills spread through to the borders with Belgium and Luxembourg and became a favourite destination for hiking. By autumn the mixed forests were a blaze of reds, yellows and browns, delighting us just as the early spring had with its brilliant greens. We'd never observed distinct seasons like this in New Zealand – if you were shown a picture of a New Zealand forest out of context, you'd be hard pressed to guess what season it was; the trees are always green. In Germany we'd now experienced all four seasons, a demarcation that showed an essential purpose to nature, a transition from one natural state to another.

The first snow fell at the beginning of December. We felt well prepared for the birth, but we could only wait. I was in the process of compiling the aircraft sampling data for publication, which I could do from our apartment to support a heavily pregnant Irena. On 19 December, Irena was admitted to Jülich Hospital with contractions. However, after a full day there was no progress with the birth and we were told she would need a caesarean. Perhaps this was the true test of my fluency in German – discussing with doctors the pros and cons of a caesarean birth over the prostrate body of my wife! Irena was put under general anaesthetic and, four minutes later, our first child, Gregory Charles, hit daylight and was most unhappy about it.

I was not allowed in the operating theatre but saw Greg a few minutes after he was born. His head looked like a pear and he was blueish pink. When I saw him, I had no feelings at all. I was in a state of shock and was petrified about Irena. She soon woke from the anaesthetic, feeling sick but cheerful. For the first couple of days after the birth, Irena and I had the same lack of empathy for our baby; we found out later this is a common reaction of many first-time parents after a difficult birth. But within a week we had both fallen in love with Greg, and Irena was breastfeeding him without trouble. With Irena and Greg still in hospital, I spent Christmas alone in our little flat, marvelling at the thought of the present we had just received.

CHAPTER 12

IRELAND AND THE ATLANTIC

When Greg arrived, all of our routines disappeared. He lost no time asserting his place in our world, the biggest casualty being our sleep. There seemed to be no end. We would drift off to sleep and then boom! The little monster would crank up the volume.

One morning early in 1980, Ulrich and I met Dieter in his office. Feet up on the desk and a Coke in one hand, he observed how tired I looked. In my sleep-deprived state I looked a total mess, with black half-circles under my eyes set off by a pallid unshaven face. Then and there, I could have easily crawled into a corner of his office and gone straight to sleep on the floor.

'Well, you like equipment. Why don't you build a gadget that responds to a baby crying and automatically rocks its cradle?' he joked, and took a swig of Coke. Ulrich suppressed a grin as we got down to business. Dieter was pleased with the formaldehyde data from the flights over the Eifel, but wanted more measurements from clean background air for comparison.

'There's an excellent site on the west coast of Ireland called Loop Head,' said Ulrich. 'It's exposed to clean onshore winds from the Atlantic Ocean and successful field campaigns for other trace gases have been run there.'

The site was well known to Dieter, and he asked us to start planning a measurement series as soon as possible.

After a frantic couple of months of preparation, we loaded the equipment into Ulrich's stationwagon and drove across Belgium and

France to Cherbourg. There we boarded an overnight ferry bound for Rosslare, on the eastern coast of Ireland. As we leaned over the railing at the stern of the ship, the lights of France fading into the distance, I felt more rested than I had in a long time. I'd dozed most of the five-hour drive to Cherbourg. And after several pints of Guinness on the ferry, Ulrich and I collapsed into our bunks. It was the best night's sleep I'd had in a long time.

The KFA had booked us into a small bed and breakfast for a week in the village of Kilkee, a couple of kilometres north of the measurement site of Loop Head. The owner was a delightful Irish lady called Mrs Enright. She welcomed us with an offer of fresh fish for dinner, caught by her husband just a couple of hours before. To set off the meal I managed to track down the only bottle of wine in Kilkee at a local pub; we were in the heart of Guinness territory and wine was considered by many of the locals to be 'foreign muck'.

The next morning we set up the sampling equipment at the brink of a cliff with a sheer 100-metre drop to the Atlantic Ocean, where waves crashed over jagged rocks. The clifftops had no trees, just unkempt grass, a few rocky outcrops, occasional clumps of gorse and peat marshes.

For the next five days my equipment ran flawlessly, and we collected about forty air samples. It was important to keep the collected air-sampling solutions cold, so I asked Mrs Enright if we could store the exposed sampling solutions in the fridge where she kept her bacon and eggs. We also needed to charge the batteries that ran the air-sampling pumps and the rotating drum sampler. Mrs Enright was thrilled to be involved, and for the rest of our visit we stored the completed sampling solutions in her fridge and charged our batteries in her garage. Because our sampling site was close by, on a couple of days she even brought us hot coffee and scones for morning and afternoon tea.

On our third morning at the site, the wind was strong, blowing directly onshore from the Atlantic. It was clear sunny weather, ideal for looking for the diurnal cycle predicted by photochemical theory modelling, which suggested formaldehyde should increase in the atmosphere after sunrise and reach a maximum at 2pm before declining until the next sunrise. Fortified by Mrs Enright's breakfast of bacon and eggs, we ran the equipment right through the day, eventually finishing at midnight. We'd hardly taken a break the whole day and, despite the bright sunshine, we were very cold. To warm up we'd taken turns running up and down a track along the cliff edge. Well after midnight we packed up and returned to Kilkee, exhausted and cold, to find that Mrs Enright had left us a plate of sandwiches and the makings for hot chocolate in the front parlour.

Over the years since, I have been on dozens of different air-sampling trips, most of which I've enjoyed. But the trip Ulrich and I had together on the Irish west coast was memorable. My equipment worked perfectly and produced data we would publish in scientific journals as well as my PhD thesis. By this time, Ulrich and I had developed a strong working relationship bound by friendship and respect. The weather on the west coast of Ireland is notoriously bad, but we had been truly lucky – to check photochemical theory, sunshine was essential and we'd had plenty of that. In fact, by the time we got back to Jülich both of us were sunburned, and a lot of jokes were made about us having just had a paid holiday in Ireland. But not the least of the success behind this campaign was the magic of Ireland, with its bleak but comforting scenery and its friendly, hospitable people exemplified by Mrs Enright.

In Jülich, Dieter was looking forward to seeing the results of the diurnal cycle that we had recorded during the sixteen-hour day. I began measuring those samples first and, although the atmospheric formaldehyde concentrations were low and obviously representative of clean air, it was clear they were remarkably uniform throughout the day and night, with no sign of an afternoon peak due to photochemistry. He was very disappointed about the results and questioned me on the reliability of the equipment. By this stage I'd built up confidence in my technique, and he soon accepted what I was showing him – that our measurements did not confirm photochemical theory for the removal of atmospheric methane by the OH radical and the role of formaldehyde as an intermediate in that process.

In science there is an adage, 'a beautiful theory destroyed by cold, hard, ugly facts'. Had my measurements just destroyed a beautiful theory? Scientists pour a lot of emotional energy into their research. When measurements destroy a beautiful theory, it's difficult not to disbelieve the results. We had to be sure there were no errors or bias in the equipment and measurements, something that required constant and dispassionate evaluation. As a scientist it's important to be aware of the limitations of both theoretical model calculations and experimental measurements. In research, modelling and experimental results go hand in hand. Finding the truth is often an iterative process, where modelling work makes a prediction that is not borne out by measurements – but contradictions with model results can often point to new experiments. Photochemical modelling had pointed to the existence of a diurnal cycle. In Ireland, Ulrich and I had not observed this. But this was by no means the end of the story.

The first autumn colours were appearing. Summer had passed in a whirlwind of parenting. Despite the sleepless nights, it wasn't long before we integrated into a threesome and became comfortable enough to take Greg with us on trips to the Eifel hills and Düren, where some of our close friends lived.

I'd been busy publishing the details of my atmospheric formaldehyde technique, as well as the results of the field campaign to Ireland, in two different scientific journals. I'd also given several oral presentations at scientific conferences in Germany, Belgium and France. Both Dieter and Ulrich were pleased with progress but felt I needed more measurements to finish what was turning into a successful PhD.

'We'd like you to undertake a major sampling campaign,' Dieter told me one morning as the three of us sat round his desk. He was in his customary pose, drinking Coke, while Ulrich and I sipped on horrible coffee. While unquestionably accurate, the Loop Head results had failed to show the diurnal cycle we'd predicted. Sampling conditions in Ireland had been perfect, with bright sunshine all day and a steady onshore wind bringing clean Atlantic ocean air to my equipment at the cliff edge. We'd sampled from before dawn till midnight. An afternoon peak in the formaldehyde data should've been there but wasn't. Dieter had been disappointed and puzzled. 'Something is clearly wrong with our current understanding of photochemistry,' he said.

They had one final sampling mission in mind: an Atlantic voyage. Ulrich explained as he threw his coffee dregs over a wilting pot plant while Dieter wasn't looking. 'We can get funding for you to sail on the *Meteor*, a German Antarctic research vessel, from Hamburg to Montevideo in Uruguay. The trip would take six weeks.'

Immediately I was torn – I didn't want to be separated from Irena and Greg for so long.

Dieter and Ulrich outlined the scientific goals. The Loop Head site in Ireland was at 53°N and the samples were taken well before the summer solstice. On a voyage across the Atlantic, the sun would be almost directly overhead in equatorial regions, with intense sunlight for photochemistry. Perhaps then my sampling would show a diurnal variation. Dieter and Ulrich looked at me intently – this was clearly something that they really wanted me to do.

After a long pause, Dieter continued. 'Dave, I've had good comments from several colleagues who've been to your presentations. After the *Meteor* voyage, I'd like you to present your work in December at the American Geophysical Union fall meeting in San Francisco.'

I was stunned. The American Geophysical Union meeting was the premier event to showcase the kind of research I was doing. It would be attended by international peers in atmospheric and climate science and had the potential to highlight not just my own work but the wider research at Dieter's atmospheric chemistry institute, and potentially attract more funding from the German federal and state governments. It was ten years since I had begun measuring atmospheric CO_2 at Makara, at a time when climate change and atmospheric chemistry were not even words on anyone's lips. The meeting would be an opportunity to learn some of the first findings of other international groups working on aspects of climate change science.

When I told Irena about the conversation, she could see it was important that I did the voyage. She said she'd invite her cousin Danuta from Warsaw to stay for a couple of weeks and would also visit Owen Rowse in London. The two visits would fill the time I was away. We also agreed that she and Greg would come to California with me in December, and maybe we could carry on to New Zealand for Christmas and New Year. Ulrich and Dieter suggested I use the opportunity to do formaldehyde sampling at Baring Head. Because this was even further south than Montevideo, the southernmost point of the upcoming *Meteor* voyage, it would be a useful supplement to the dataset from the ship, extending the geographical range of the measurements.

A couple of days later I had a very welcome visit from Dave Keeling, who had been on sabbatical at an atmospheric science institute in Berne, Switzerland.

'G'day Dave,' he drawled, in an imitation of a New Zealand accent, as he poked his head round the door of the lab. 'I couldn't leave Europe without dropping in.'

Keeling knew Dieter Ehhalt well and had been keen to visit the atmospheric chemistry institute as well as me.

'Dieter tells me you have a presentation planned for the American Geophysical Union meeting. I'll also be there, but how about dropping by Scripps on your way? I'm keen to talk about the Baring Head CO_2 programme.'

Since I'd left New Zealand, Martin Manning with other staff from the INS had been running the Baring Head programme successfully. Keeling was keen to use the measurements to help show the role of El Niño and La Niña events on the large-scale absorption of atmospheric CO_2 in the Southern Ocean and Pacific region more generally. It would be a serious undertaking, requiring cooperation between international labs in several countries, including New Zealand, Australia and the US, and he wanted my thoughts on his ideas.

*

The ship rolled, corkscrewing a second later as the bow buried itself with a violent shudder in a trough between huge waves. We'd sailed down the Elbe to the northern German coast where a force 11 storm was raging. For context, force 12 is considered a hurricane. The captain had decided to wait out the storm at anchor for a day before venturing into the English Channel at a force 9 gale. The *Meteor*, a relatively small ship, rolled and pitched in all directions. Most of the scientific staff on board were violently seasick, me included. The lab had a large sink in one corner into which I periodically vomited until my stomach was empty. This was far worse than my trip from Wellington to Antarctica a few years before. There I'd got used to the rolling and my body had adjusted to walking along a moving deck, my stomach behaving itself. Here the movement was unpredictable, and you could be flung into a bulkhead. The ship's doctor, also terribly seasick, was treating a lot of injuries.

Before leaving, Ulrich and I had spent three days installing the equipment on the ship. The rotating glass sampler was mounted on a boom facing forward over a bulkhead above the *Meteor's* bridge, directly exposed to air coming off the sea from the bow. The analysis gear, including a portable HPLC that I had purpose-built for the trip, was set up in a small lab at the stern and shared with an oceanographer from the University of Kiel in northern Germany. Collecting air samples on the top deck during the storm was out of the question – waves were breaking over the bow of the ship and, although my sampler was in the clear, it was far too dangerous to go out onto the decks. Although I managed to get some preparation work done in the lab, I spent most of that day and night in a foetal position on my bunk knowing that I wanted to die.

The ship's captain explained that the *Meteor* was being subjected to 'confused seas' in the English Channel; waves were coming at us from all directions. 'Don't worry,' he told us, 'it will get better in the open waters of the Atlantic. We'll probably get bigger waves but they should all be coming from one direction and easier for you landlubbers to cope with.' He jovially tucked into a large roast dinner with a dozen ashen-faced scientists and students looking on. The ship was on its way to Antarctica, and most of the research on board involved marine biology and oceanographic experiments covering both the northern and southern Atlantic. However, there were a few other atmospheric experiments aimed at measuring hydrocarbons as well as gases containing nitrogen. Unfortunately the science leader on the voyage was a bully and, although I had no issues with him, several of the other scientists and students had to put up with mistreatment. I've met a few science managers like this in my scientific career, and invariably their science suffers – the most successful research is built on good teamwork, and above all is driven by a leader who encourages staff and their research rather than putting them down.

After a miserable night, I managed to keep down a weak cup of black tea and a dry bread roll. The ship's rolling had subsided a little, or I was starting to adjust to the constant motion. I started air sampling on the top deck, managing to take four samples that day, all of which I measured on the HPLC. It was unpleasant and cold on deck, but being in the fresh air was definitely the best part of the job. Far worse was having to analyse the samples in the confined, heaving space of the lab. Invariably I would throw up in the corner sink between analytical runs.

My equipment was working well but the data were quite variable. This was not surprising because we were in the English Channel and surrounded by ships; I counted at least twenty that day, including four giant oil tankers. We were also close to land and I could clearly see the white cliffs of Dover on one side of the ship and Calais on the other. Pollution sources on both sides of the channel, as well as from the ships, were definitely affecting my measurements.

Over the next few days, the strong winds and heavy seas continued, but perhaps I was developing sea legs because I was able to increase the number of samples I was taking. The conditions continued to be utterly miserable; on the top deck where the boom holding my sampler was mounted, wind speeds were often over 30 metres per second. I had to shuffle along the deck, holding the railings, to avoid being blown over. A couple of minutes outside and my hands and body would be numb with cold, and it would take me an hour in the lab to warm up. I was settling into an exhausting routine, taking the first sample at six or seven in the morning and ending at eleven at night. Each hour I tried to take a fifteen-minute break between sample analyses to talk with the others onboard. I was still seasick, but the ship's doctor had given me a packet of ginger tablets that some of the others had found useful.

I missed Irena and Greg terribly and looked forward to weekly radiotelephone calls, when Ulrich would relay news of them and advice from Dieter. The calls were routed from the German telephone system via Norddeich Radio in the north of Germany to the *Meteor*'s radio room, and I got to know the ship's radio operator well. My father had told me about his adventures as a young radio operator on merchant ships in the Atlantic in the 1930s. Talking to the *Meteor* radio operator about his experiences on various oceanographic voyages and his equipment gave me respect for what my father had achieved aboard the ships he'd been on. I was not prepared for the feelings I had in the *Meteor*'s radio room, moved to tears thinking of the precious years I had wasted with my father after my marriage breakup, when I could've reached out to him. I resolved to stay in constant communication with my own son Greg, no matter what.

Ten days after leaving Hamburg, despite the punishing sixteen-hour days, I was starting to enjoy the voyage. We were well out into the Atlantic Ocean and had passed the Canary Islands, about 100 kilometres to the east. We'd seen no land since Dover and were not expecting to for the next two weeks, when we would reach the northern coast of Brazil. The sea was calmer and azure blue, with our wake stretching to the horizon. It was getting warm and I changed into shorts and sandals.

'Herr Lowe, get out!' shouted the captain as I entered the ship's mess at lunchtime. Puzzled and not a little embarrassed, I backed out of the mess.

'Captain Feldmann has a very strict dress code,' one of the German scientists told me, 'and will not allow anyone to wear sandals in the mess. It's a real drag in hot weather, but on this ship, he's the boss.'

After that, I would change into long trousers for meals. As we got closer to the equator the trousers became unbearable, especially in the lab. Eventually I settled on socks and sandals with my best shorts and, although this raised eyebrows in the mess, I was not shouted at by Captain Feldmann. Following my lead, most of the other German scientists and students started wearing shorts, socks and sandals into the mess – they figured that if the New Zealander could get away with it, then they could too.

By the time we'd reached the mid-Atlantic, my formaldehyde measurements were showing much lower levels, indicative of clean air, and scientists on the ship measuring other atmospheric trace gases were also measuring very low values. We began seeing remarkable red skies with layers of sand appearing on the decks – this was dust blowing in from the Sahara Desert, over 1000 kilometres to the east.

'It sounds like very clean air,' Ulrich said one day on his weekly phone call. 'Do you think now you could repeat the diurnal experiment you and I were unsuccessful with in Ireland?'

'I'll have to work right through the night, but I think I could manage that.'

'Actually,' started Ulrich, sounding a bit embarrassed, 'Dieter would like you to run the experiment for two consecutive days so you get forty-eight hours of measurements. He thinks then you should be able to see the formaldehyde concentration is higher during the day.'

Their suggestion made sense. The sun was higher than it had been in Ireland, and this was the time to try the experiment.

After two days and nights with very little sleep I had the results. There was no diurnal signal. The data were a rerun of Ireland earlier in the year. The formaldehyde concentration remained uniform and low, despite bright sunshine during the daylight hours.

'Are you sure?' asked a disappointed Ulrich when he phoned a couple of days later.

'Yes, absolutely. But we're still more than 30° N and the sun is in the southern hemisphere. How about I repeat the experiment around the equator, with the sun almost directly overhead?'

'That's a great idea,' said Ulrich. 'And by the way, I just spoke with Irena – she and little Gregory are fine and send you their love.'

What a vital link that radiotelephone was. I was missing Irena and Greg terribly, but focusing on the work day and night meant that the trip was passing quickly. Just north of the equator, the sea and air temperature had risen to 28°C, and working in the jammed space of the lab below the decks was almost unbearable. Just before sunset each night, a dozen of us would go up to the top deck to try and spot the 'green flash' as the sun went below the horizon, the same elusive refraction of light that I'd looked for every Friday evening over the beach at Scripps in California. One evening on the ship all of us saw it – it was unmistakeable. One moment the sun was a crimson orb sliding towards the horizon, but suddenly, as it disappeared, the red was replaced by a lurid green flash lasting for a second or so. Then the sun was gone and darkness fell within minutes. Close to the equator there is little or no twilight; sunset is like a light being switched off.

I repeated the diurnal experiment over a four-day period as the ship sailed across the equator from 8°N to 8°S. There was no way I could stay awake for the full four days so I compromised, running the air sampler for longer periods, up to three hours, which meant that I could snatch about an hour's sleep between samples. When I started measuring the samples in the lab, a weak diurnal cycle appeared with a small maximum at around midday on all four days, especially on the day when the sun was directly overhead. The results still did not match the photochemical models; the maximum was much smaller than predicted, but it was definitely there. I had provided direct evidence that photochemistry was actively involved in the removal of methane from the atmosphere. It was a world first and a highlight of my PhD thesis. Later, other researchers corroborated these results and expanded them to include further aspects of photochemical theory, all of which now form an essential part of the field of atmospheric chemistry and its role in climate change science. Ulrich and I went on to publish the results in a widely cited paper in the *Journal of Geophysical Research*, one of the first to focus on research aimed at understanding human modification of the planet.[11]

After a couple of days sailing south of the equator, we spotted the broad sandy beaches of northern Brazil, backed by coconut palms and jungle. We docked in Recife, about 400 kilometres southeast of the mouth of the Amazon River. I lost no time getting off the ship with Rudinger Rasmus, one the German crew members. We delighted in walking on dry land along wide, palm-lined streets of beautiful colonial Portuguese-style buildings. Our first stop was for a freshly brewed Brazilian beer with the unlikely name 'Antarctica'.

German friends had recommended visiting Caruaru, a town about 150 kilometres inland. Brazilians speak Portuguese, but I managed to communicate in my limited Spanish and bought bus tickets to Caruaru. The landscape changed rapidly from tropical jungle to inland desert, with prickly pear and stunted thorn trees. Caruaru was fabulous, and spacious, with parks and churches. There were no other tourists around and two men began competing to be our guide. Unfortunately this turned into a fistfight, so we retreated into a bar for another Antarctica and eventually got back to Recife about midnight. Boarding the ship involved running a gauntlet of prostitutes on the dock, beautiful young women who tried to fondle us as we walked by. As we sailed south to Rio de Janeiro over the next few days, the ship's doctor was busy treating the crew and some of the scientific staff for various sexually transmitted diseases.

I continued working exhausting days running the formaldehyde sampler as we sailed south. The data began to show a lot of scatter due to multiple pollution sources, including fires all along the coast. I took two samples as we entered the harbour of Rio de Janeiro and these were the most contaminated I'd ever come across. There was no point in running any more experiments, so with a sigh of relief I sat on the deck with Rudinger and shared a bottle of rum as we sailed into the most beautiful harbour I have ever seen.

'Hey, check that out,' said Rudinger, handing me a pair of binoculars.

At the top of Corcovado hill towered the 50-metre statue of Christ the Redeemer, arms open wide. It was an emotional experience. My father had sailed into this harbour on a merchant ship almost fifty years before. He had told me it was the most beautiful harbour he'd ever seen. The sun set and darkness fell swiftly over the bay. The Christ statue shone like a beacon, an awe-inspiring sight.

The next day I was hungover and feeling seasick, thanks to the dubious bottle of rum that Rudinger and I had finished the night before. A violent squall was pounding the ship. One of the crew had seen a twister as the sun rose. The magic city of Rio de Janeiro was gone.

My head soon cleared when I began collecting and analysing air samples on the final push to Montevideo, the planned endpoint of my trip. Montevideo is on the northern bank of the Río de la Plata, or River Plate. The river mouth is so wide it's impossible to see the southern bank, Argentina. We dropped anchor in the port only to spend more than two days cooped up on the *Meteor*, a couple of hundred metres from shore. On the first evening the science leader, the captain and two senior ship officers went ashore on a small boat and stayed overnight at an official reception held in honour of the research voyage, leaving the rest of us behind. The mood was ugly when the shore party returned the next day: a series of expletive-heavy notes had been taped by the gangplank and the party was jostled and booed as they came aboard. The science leader had continued his obnoxious tactics and, for many researchers onboard, the trip couldn't come to an end soon enough.

The trip back to Germany from Montevideo was a forty-hour nightmare muddle of delays and missed flights. From Montevideo to Porto Alegre, São Paulo, Rio de Janeiro and Recife, then over the Atlantic to Faro and Lisbon in Portugal. Stuck there for ten hours, I finally made a flight to Frankfurt then on to Düsseldorf and a reunion with Irena, Greg and Ulrich.

The *Meteor* trip had been gruelling at times but an outstanding scientific success. I was proud of what I had achieved. I'd shown that formaldehyde measurements made in clean air could be used to understand the natural processes that remove methane from the atmosphere. Since then, measurements of atmospheric methane made at many international sampling stations, including Baring Head, show that its concentration is growing rapidly and exacerbating the already serious climate situation. Methane is produced by both industrial and agricultural sources and, because of its links to food production, has become a highly charged political issue. In New Zealand, for instance, a relatively significant proportion of total carbon emissions are due to methane from ruminant animals, cattle and sheep. Because of its importance to exports and food production, the agricultural sector has so far not been included in the New Zealand emissions trading scheme, a decision seen as unfair by other carbon-emitting industries and groups advocating for reductions in carbon emissions. Measurements of methane worldwide will be an essential basis of any future political decisions on the urgency of emissions reductions.

PART IV

1980–2007

337–381 ppm

Accelerator target prep for radiocarbon dating

INS/DSIR 1984

CO_2 reduction reaction

$$CO_2 + 2H_2 \xrightarrow[Fe]{550-800°C} C + 2H_2O$$

goes via several steps incl. CO

but at $T > 800°C$ "producer gas" reaction

$$H_2O + C \xrightarrow{800-1100°C} CO + H_2$$

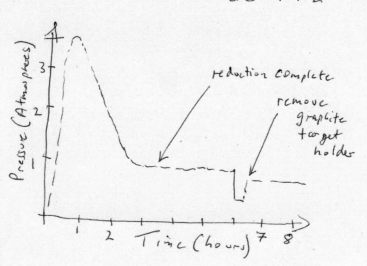

reduction complete

remove graphite target holder

graphite deposited

accelerator ion source target holder

CHAPTER 13

THE COUNTRY THAT CHANGED US FOREVER

I looked at the group of scientists at the back of the packed meeting room – Dave Keeling and Peter Guenther were clapping, along with the rest of the audience, which included Dieter. I'd just finished detailing the formaldehyde results from my aircraft and field campaigns. I was embarrassed and humbled. The American Geophysical Union's fall meeting was the foremost in the world for presenting research in atmospheric and oceanographic sciences.

There were a lot of questions, which I took to be a good sign. The first was about measurements aboard the *Meteor*, and I described the portable HPLC I had built and the flexibility it had given me to adjust sampling as I crossed the Atlantic. Most of the other queries covered various aspects of the technique, methane photochemistry and the results of the flights over the Eifel hills. Eventually the chairman moved on to the next session. I sat down, drained but elated. I'd never spoken to such a big group before let alone at such a prestigious conference. At the next coffee break Dieter congratulated me – he'd already had favourable comments from his international colleagues, and my presentation would help promote the work of the atmospheric chemistry institute in Germany.

Nowadays, an interdisciplinary approach to tackling human-induced climate change through international panels and meetings is standard practice, but those inspirational sessions in San Francisco 1980 marked the first time I can remember experts from so many different areas of science gathered together. I met ecologists, social scientists and economists as well as atmospheric scientists, all focused on the likely repercussions of increasing atmospheric CO_2 on the planet and its life-sustaining environment.

As planned, Irena, Greg and I travelled back to New Zealand for Christmas and New Year. 'Some things never change,' I thought in the roaring wind at Baring Head, as I set up my equipment near the cliff edge after more than two years away. I was running a full day and night sampling for formaldehyde, and was settling in for a blustery night with a borrowed sleeping bag and a pack of sandwiches when Steve the lighthouse keeper arrived. He offered me a bed at his place and an alarm clock so I could wake and change my samples every hour or two through the night – I gratefully accepted.

The sampling equipment worked well, and a day later I managed to measure the samples on a borrowed HPLC based at a DSIR lab in Lower Hutt. Because I'd made the measurements only three weeks after the southern summer solstice in New Zealand, Dieter Ehhalt expected that I would record a maximum in formaldehyde in the afternoon. But the results were similar to the data measured in Ireland a year earlier, with no significant maximum during the day, despite bright summer sunshine. Once again, my measurements had contradicted current photochemical models projecting the removal of methane from the atmosphere. Only on the *Meteor*, near the equator, had I detected weak maxima that could be attributed to photochemistry. It would be many years before this particular puzzle would be solved.

Our couple of weeks in that early New Zealand summer was a delight. This was the first time our parents had met Greg; it was a proud and emotional time. We made the most of the short summer break knowing we'd soon be returning to winter cold.

In Jülich, temperatures had been well below freezing for a month. But happily, the fields were covered with snow, the ponds and lakes were frozen, and there was bright sunshine. Irena borrowed ice skates and started skating on the frozen lake in the middle of Jülich. The bright winter scenes were exhilarating, a complete contrast to our first depressing winter.

Ulrich and Dieter had one more mission in mind for me before I could complete my thesis. Back in Dieter's office one morning – his feet up on the desk, as usual – they asked me how I felt about doing one more sampling flight over the Eifel hills.

'Last time you got up to an altitude of 5 kilometres,' explained Dieter. 'We'd like you to sample quite a bit higher, up to seven kilometres. Looking at the photochemical models, we think that the formaldehyde concentration should drop off quite dramatically as you get higher above the inversion layer.'

Ulrich and I got to work installing the equipment in the rear of a modified business aircraft, a Piper Navajo, which would fly out of a small airfield near Aachen, Merzbrück. The air intake was mounted outside, forward of the pilot, and a short test flight satisfied us that everything worked perfectly.

Next day we climbed to 4 kilometres above the Eifel, and I began sampling. Above 5 kilometres the pilot asked me to put on an oxygen mask and headphones so that I could hear his instructions, and I noticed that I could also hear external radio communication with our aircraft. The oxygen mask was bulky, and I had to take it off every few minutes to run the equipment. The effect on my brain was almost immediate, with a noticeable slowing of my ability to make decisions. Remarkably, one puff of oxygen every few breaths was enough to get me thinking clearly.

Things were going well. I'd collected most of the sample at 7 kilometres when I heard an urgent announcement over the aircraft radio.

'Warning to unknown aircraft over the northern Eifel. You are in a shooting area – get out immediately!'

A NATO jet buzzed close by and the pilot began shouting that we had to descend, nosing the aircraft down rapidly to an altitude of 5 kilometres. As he turned back to our airfield, he grinned at me from the cockpit. 'That was exciting!' he said.

During the high-altitude flight, I'd managed to collect five samples showing very low formaldehyde concentrations above the inversion layer, consistent with photochemical models. It was an excellent dataset and was the last aircraft sampling I required. Now it was time to focus on writing up my thesis, completing lab tests and preparing for my oral exams.

Because the atmospheric chemistry institute wanted to publish my thesis internationally, the university allowed me to write it in English. But the oral exams – three *Nebenfächer*, or unrelated subsidiary subjects, and one on my thesis – were to be in German. I wasn't bothered by the thought of the oral examinations but I was concerned by my timetable. Ulrich had insisted I complete a series of control experiments as a final test of the atmospheric formaldehyde technique. In addition, I had to write my thesis and have it ready for submission, take three different lecture courses for the *Nebenfächer*, and pass the oral exams. And be a husband and father as well!

For the three *Nebenfächer* I chose nuclear chemistry, low temperature physics and electronics. Because the university had a close affiliation with the KFA, many of their science professors were based at the research centre, and this included all three professors lecturing the courses I had chosen. The professor taking the low temperature physics course amazed me by telling me I should prepare for his exam by reading recent copies of *Physics Today*, a well-known international journal; he did not want me to go into Cologne to attend lectures, which was a huge saving in time.

For the nuclear chemistry course, I had to attend lectures at the university. However, the professor organised a chauffeur from the KFA to drive us and two other students to the university once a week for the six-week course. The chauffeur would often drive the Mercedes at over 200 kilometres per hour on the Autobahn, overtaking lines of heavy trucks; my knuckles would be white, but the others barely noticed the speed. For the remaining course in electronics, I attended only a couple of labs before the professor told me that I obviously had enough experience in the field already and needn't return to classes before the exam.

The PhD thesis was going to cost 3000 Deutschmarks to be

professionally typed – a vast sum for a student. To make things worse, Dieter Ehhalt refused to read my handwritten drafts. Our German friends were unsympathetic. 'Maybe you should learn to type,' they joked. So, I bought myself a small portable electric typewriter and did just that.

The days, weeks and months blended into one another. Greg was getting more active, and we were taking him everywhere – on our bikes, or shopping and walking with a stroller. He seemed interested in everything he saw, especially aircraft flying overhead. Usually these were NATO jets flying out of nearby US and British military airfields, a hangover from the allied occupation following World War Two.

Watching our son change from a baby into a toddler, becoming aware of his environment and sharing life with us in Germany, was a special experience; by this stage we were quite fluent in German and would discuss politics, the economy and life in general with our German friends. Greg was spoken to in German by our friends and it was hardly surprising that his first word was '*Doch!*', which loosely translates into New Zealand English as 'That's for bloody sure!' We decided the timing was right for another baby, due in January 1982.

Although Jülich was not far from the beautiful Eifel hills, our bike rides around the immediate landscape were anything but picturesque. The backdrop to flat fields of sugar beets was coal-fired power plants with towering plumes filling the atmosphere with pollutants. Even on a fine day the air was hazy, especially towards the horizon, which was usually a washed-out whitish grey. The plants were fed by brown coal from enormous open-cast mines, at least a dozen within 20 kilometres of Jülich. The closest mine was Hambach, which is still operating today. With an area of 85 square kilometres, it's the largest mine in Germany. We would often ride our bikes, with Greg in a baby's seat, to an observation platform overlooking the pit. The size was staggering; supposedly the giant hole was visible from the moon with a pair of binoculars.

The brown coal is mined by giant continuous tracked machines called 'baggers', almost 100 metres high and more than 200 metres long, weighing around 14,000 tonnes each. These are run on electricity from power stations fired by the very coal they dig up, transported to the station's furnaces by enormous conveyor belts also powered by electric motors. If you think there is something circular about this, you're right. It's a dirty and inefficient way of producing electricity, with many terrible side effects; whole villages and forests have been erased to make way for the mines. The massive hole at Hambach drains groundwater from the entire region, resulting in 'mine quakes' where houses suddenly shake, subside and shift sideways. And, not least, the power plants make a major contribution to atmospheric CO_2 – in many countries, coal-fired power plants are the single biggest contributor to CO_2.

The atmospheric chemistry institute was based at the KFA nuclear research centre, where two experimental nuclear reactors were a primary focus. As a way of promoting the benefits of nuclear power, the management touted it as being CO_2 emissions-free. This was a tightrope that Dieter Ehhalt had to negotiate, because all of the research at his institute was environmental. The risks of nuclear power were already well known, even before the catastrophic accident at Chernobyl in the Ukraine in 1986, and he did not want to be seen actively promoting nuclear power.

The early 1980s marked the beginning of huge political and industrial arguments over both brown coal and nuclear electricity production in Germany, with ugly battles between protestors, the police and miners. The KFA was targeted. By the end of the 1980s, the KFA's name – Kernforschungsanlage – was changed to Forschungszentrum Jülich, meaning Jülich Research Centre. For political reasons the word *Kern*, or 'nuclear', was removed from the title.

In Germany we saw at first hand the interplay between politics, the fossil fuel companies and the energy generators, including the nuclear industry. This formed a backdrop to my study into atmospheric chemistry. The enormous social and environmental cost to the German public weighed on my mind, and visions of the forests and villages that simply disappeared off the map when the gigantic hole at Hambach was created have been with me ever since.

A couple of years after we returned to New Zealand, the nuclear catastrophe at Chernobyl in 1986 hugely impacted the mindsets of Germans and other Europeans. After the explosion, a radioactive cloud spread north and west from the stricken Ukrainian reactor, depositing radioactive material over large parts of Germany, contaminating land and waterways. Measurements showed that eating large-leafed vegetables like lettuce and cabbage was potentially dangerous, but the health and risk information coming from German authorities was often confusing and contradictory. Sceptical of the government's advice, friends in Jülich used a Geiger counter on the soles of their shoes to measure radioactivity picked up from walking outside on contaminated land. The levels were often so high that they chose to leave their shoes outside the house, walking inside in socks and slippers. The country went to huge lengths to deal with the contamination, including destroying crops, cleaning cars that crossed the borders and replacing sand in children's sandpits.

The disaster proved a pivotal point for the German anti-nuclear movement, with clashes between protestors and police increasing. The stakes were high, because Germany had invested a significant amount in nuclear power production and its leaders were concerned that the Chernobyl disaster could lead to non-acceptance of this programme by the public. And yet nothing was to change until in 2011, when Japan's Fukushima nuclear disaster marked the beginning of the end for the nuclear power industry in Germany. Hundreds of thousands of protestors demonstrated throughout the country, and within months the government made the decision to shut down all nuclear power plants by 2022 and begin a transition to renewable energy.

A beaming Dieter Ehhalt was congratulating me on completing my fourth and final oral exam and having handed in my PhD thesis. I had passed with a *magna cum laude*, or a grade A, which had never been achieved by a foreigner at the KFA before.

'I was especially pleased,' Dieter continued as we sat in his office, 'to get a phone call from your physics professor this morning, who told me that you should be studying with him rather than wasting your time on atmospheric chemistry. Let me tell you, Dave, it's a great feeling when another professor says that my students actually know something!'

At that moment there was a commotion outside his office door. There was Ulrich, carrying a huge bottle of *Sekt*. The other students and staff picked me up and carried me through to the tearoom, where a lunch had been set up. Later that afternoon I wobbled my way home on the bike, collapsed in a chair and fell asleep.

The formal graduation ceremony was held at the university the following week. Irena and Greg were there, as well as several of our German friends and other students from the atmospheric chemistry institute. A tradition in Germany has graduates presented with a so-called 'doctor's hat' designed around the theme of their thesis. Ulrich had arranged for the institute's workshop to build an enormous hat covered with motors and flashing lights and a rotating plastic drum full of Raschig rings. It was so heavy I nearly fell over!

I still had six months of my New Zealand study award to complete, and so I continued to work as a scientist at the institute. I helped supervise a new PhD student, worked on international publications covering my thesis results, and helped some of the students and staff prepare for a conference in the US.

Our baby was due in mid-January 1982 and we were not sure how we would cope. Irena's mother came to the rescue. She arrived direct from a New Zealand summer but she loved the snow, telling us it reminded her of Poland when she was a girl. She even rode Greg's sled down a small hill, much to his delight.

Irena's contractions began on 7 January. I drove her 30 kilometres to Aachen Hospital, and within an hour Suzanne was born, a wonderful healthy child yelling her head off. After a sleepless night at the hospital I took the Autobahn back to Jülich, gasping at the cold as it started to snow. It snowed heavily for two days, and with all the roads and trainlines blocked there was no way of getting back to Aachen. The temperature dropped to -20°C and the doors and windows of the car were frozen solid on the inside; starting the engine let alone driving the car anywhere was impossible.

By the time Irena and Suzanne were discharged, the weather had warmed up and I was able to bring them home. Imagine the four of us and Irena's mother cooped up in our tiny apartment – it was a madhouse, with little time for sleep. Greg had to adjust. He soon learned that kicking his tiny sister went down badly with us and he settled into a kind of simmering indifference.

The bitter winter turned into spring. A passionate cyclist, by this point I'd biked every forest and field path within 20 kilometres of the town. Michael Trainer, one of the students at the institute, suggested a more ambitious ride. We would cycle from Jülich to Brussels, a distance of almost 200 kilometres in one day, returning by train. We set off early on a beautiful Saturday morning through Germany, across the border into the Netherlands, where we headed for Maastricht and into Belgium without seeing the border. We headed straight for Brussels, arriving exhausted but exhilarated at about 8pm. On that trip I noticed things I'd never have seen in a car. Cycle touring is a fabulous way of seeing and feeling a country, and with its variety of cultures and nationalities and safe bike paths, Europe is a wonderful place to cycle. I've ridden bikes all my life on holiday and to work in three different countries. In the twenty-first century, the advent of electric bikes has dramatically increased the numbers of people cycling to work and for shopping and recreation. Unfortunately, in many countries, including New Zealand, the provision of dedicated cycle paths has not kept up with the growth in cycling, leading to serious accidents. Providing cycle paths is an obvious way of reducing commuter traffic, carbon emissions and improving health.

*

At the end of May 1982 it was time to return to New Zealand. Ulrich drove the four of us to Brussels Airport and after we'd checked in, he turned to me. I had tears in my eyes as we hugged goodbye. 'We'll see each other again,' he promised.

In Germany we'd experienced four winters and the seasons between, learning to appreciate the birth of spring growth, summer and the natural passage through a colourful autumn into the short dark days of winter. We'd come to know, understand and appreciate Germany and its people, language and culture. We felt different, completely enriched by our experience, knowing that our future would always be tied to Germany and the lifelong friendships we had made there. I'm very much a New Zealander but I will always feel part German.

I knew too that what I'd learned would help me, with other scientists in New Zealand, establish an international programme in atmospheric chemistry based at Baring Head. I was certain that Germany, the US and other countries would have a major role in our research, and that my work and study to date had prepared me for what was to come. The atmosphere swathes the entire planet. It has no boundaries and, to understand it, international scientists and leaders need to look past boundaries and differences too. Baring Head was to be part of this.

CHAPTER 14

FINGERPRINTING ATMOSPHERIC CARBON

We'd left New Zealand as a couple and arrived back as a family of four with a completely different outlook on life. Everything was familiar but at the same time felt strange and uncomfortable. It was as though we were foreigners in our own land. Settling back into our old lives turned out to be difficult. We both felt homesick for what we had left behind in Jülich and the deep friendships we had made there. It was difficult to relate our experiences in Germany to friends who had never lived and worked abroad and could only answer our pain with glazed expressions.

Adding to this was the economic and political turmoil that mired New Zealand under the leadership of Robert Muldoon's National government. The country had stagnated with inflation in high double digits, growing unemployment, high external debt and borrowing from overseas markets. Everything was expensive and prices of goods seemed to increase every week. The New Zealand we'd arrived back to was one of disastrous economic policies and had just the year before been socially riven by unprecedented civil disorder over the Springbok Tour.

It was mid-winter, nowhere near as cold as the German winters, but the same southerly winds that provided perfect conditions for sampling CO_2 at Baring Head brought cold, wet storms to the Wellington region. Most New Zealand houses have famously poor insulation, with gaps under doors and windows, and temperatures inside the houses often barely exceeded the outside temperature. The old house in Lower Hutt was no exception. On clear cold winter mornings after a storm had passed through, we could see our breath condensing before our eyes, and the windows in one of the bedrooms routinely had ice on the inside. The house had seemed fine when just Irena and I lived in it, and it was bigger than our apartment in Jülich, but with two children constantly underfoot, the place seemed to have shrunk. It was obvious we were going to have to extend the house and fully insulate it but it was heavily mortgaged with crippling repayments due to high interest rates. We also needed a car, new appliances and clothes, but everything seemed to be more expensive and of poorer quality than in Germany. In those early months of our return, money disappeared like water.

Going back to work at the Institute of Nuclear Sciences was also a shock. Athol Rafter, my original employer and mentor, had retired as director. His successor, Bernie O'Brien, congratulated me on my PhD and welcomed me back to the institute. However, he did not want to expand the institute's atmospheric chemistry research any further than the CO_2 programme at Baring Head. On my first day back at work he made it clear that the INS needed to stick to their name and couldn't expand into areas that didn't involve nuclear or isotopic research.

I was bitterly disappointed – my head was full of ideas I'd brought back from Jülich, and the thought of limiting important work to one set of scientific tools seemed an unnatural constraint. When I spoke about the situation to Martin Manning, who was successfully running the Baring Head station, he sympathised. He told me how Bernie was constrained by the bureaucrats at the DSIR, and suggested that if I came up with a good idea for atmospheric research which used the techniques available at INS, he was sure they would back me.

A large part of the research at the INS had been founded on the pioneering work of Athol Rafter in the field of radiocarbon dating in the 1950s. In the 1960s, Athol's work had looked at radiocarbon dating in both archaeological and geophysical studies. This included measurements of radiocarbon in atmospheric CO_2 at Makara as well as carbonates dissolved in ocean waters around New Zealand. To understand the significance of this work and its impact on the future of atmospheric chemistry and climate change research, we must start with a few details on the role of carbon isotopes – meaning variations of carbon with different numbers of protons and neutrons in their nuclei – including carbon-14.

Naturally occurring carbon in the biosphere, atmosphere and oceans contains three different isotopes. Two of these are stable: carbon-12 and carbon-13. Carbon-12 is the most common, forming about 99% of all carbon in nature, whereas carbon-13 makes up only about 1%. The third carbon isotope – radiocarbon, or carbon-14 – is present in vanishingly small quantities, about one atom in every trillion carbon atoms in living things. Carbon-14 is produced in the upper atmosphere, when high-energy cosmic rays from outside the solar system bombard nitrogen molecules in air, resulting in a cascade of nuclear reactions. Neutrons (neutral subatomic particles) produced in the cascade react with atmospheric nitrogen to produce carbon-14, a naturally radioactive form of carbon with a relatively long half-life, meaning that in this case it takes 5700 years for half of it to decay back to the parent stable isotope of nitrogen, nitrogen-14.

After carbon-14 is produced, it becomes part of Earth's carbon cycle through its integration into atmospheric CO_2. During photosynthesis, green plants absorb CO_2, converting the carbon into sugars; this is the start of the food chain for living things grazing on plants – including humans, directly from eating vegetables and fruits, or indirectly from drinking milk and eating products derived from animals that have fed on plants. In this way, the carbon-14 that was originally in atmospheric CO_2 becomes part of the carbon contained in our bodies. Hence, all living species that feed directly or indirectly on plants contain carbon-14 and are naturally radioactive. While a particular species is alive, the carbon-14 it contains remains at a relatively constant level set by the production of carbon-14 in the upper atmosphere. However, after the species dies, exchange with carbon-14 in atmospheric CO_2 ceases immediately, and during the next 5700 years, half of the carbon-14 decays back to stable nitrogen-14. This means, for example, that a 5000-year-old Egyptian mummy will only have about half the amount of carbon-14 that a living person today has.

About seventy years ago, Willard Libby, an American physical chemist, demonstrated that the carbon-14 content of carbon-containing material, after suitable sample preparation, could be measured using modified Geiger counters. The result was a revolution in the field of archaeology. Because living things stop absorbing carbon-14 at the time of their death, measuring the carbon-14 content of 'dead' artefacts and species provides a way of determining the age of human and other relics that were originally deriving their carbon from the air.

Athol Rafter visited and worked with Willard Libby in the 1950s and brought the technique back to New Zealand. He used equipment installed in the Shed 2 laboratories at the INS to provide a vital service to New Zealand archaeologists. But he also knew that the technique would be a valuable way of detecting changes in atmospheric CO_2 itself.

From measurements at Baring Head, Makara and elsewhere, we knew that CO_2 was increasing in the atmosphere due to the combustion of fossil fuels like coal, oil and natural gas. Fossil fuels are formed when ancient organic matter is buried deep in the earth and subjected to heat and pressure during hundreds of millions of years. The timescales since this organic matter was alive are so long that all of the carbon-14 originally present disappears, and the material is carbon-14 free or 'radiocarbon dead'. When these fuels are burned in power stations, motor vehicles and elsewhere, the CO_2 released into the atmosphere is also carbon-14 free. In a series of measurements in the 1950s, Athol Rafter was able to show that this was actually decreasing the carbon-14 radioactivity naturally present in atmospheric CO_2. It was, potentially, 'smoking gun' evidence proving that the release of CO_2 from the combustion of fossil fuels was causing atmospheric CO_2 to increase, with implications for climate change.

If only life and the atmosphere were that simple. In the 1950s and 1960s, the US and the Soviet Union (and to a lesser extent, the UK) were testing nuclear weapons in the atmosphere; hundreds of blasts were set off in the atmosphere of both hemispheres. These bursts generated carbon-14 in the atmosphere through the production of neutrons, in a way similar to the natural process in the upper atmosphere with cosmic rays. The impact of the nuclear weapons testing was dramatic: by the mid-1960s, levels of carbon-14 in the atmosphere had risen to twice the natural levels observed before the advent of the weapons. Atmospheric carbon-14 data published by Athol Rafter and others showed an alarming and exponentially increasing spike in the record. In 1963 a limited test ban was signed by the US, Soviet Union and UK, and most testing was transferred underground.

This was the height of the Cold War, and there are many rumours as to why the United States and Soviet Union signed this agreement – there was little else they agreed on at that time. One theory had to do with the dramatic increase in radioactivity in all living things due to carbon-14. If this continued, what would be the effect on living things? Would carbon-14 levels in food become so high that it would be dangerous for humans? Fortunately, we will never know. My own opinion is that it was clear that testing colossal hydrogen warheads in the atmosphere was madness, with all sorts of unknown consequences for the environment and humans. In many cases, the same weapons test data could be obtained from underground blasts and laboratory simulations.

Athol Rafter was the first to demonstrate the exponential increase of what came to be known as 'bomb carbon-14' in the atmosphere of the southern hemisphere. Rumour has it that, in the early 1960s on a visit to the US, where he was due to present his results at a scientific meeting, he was taken aside by the CIA at Los Angeles Airport and told not to show the data. Regardless, he eventually published his findings in scientific journals, and is widely credited as one of the discoverers of the 'bomb effect'. Ironically, the bomb effect – with its huge transient spike in carbon-14 – has proved to be a useful tracer in geophysics, for example when investigating the uptake of atmospheric CO_2 in the oceans.

So, my introduction to the INS had included radiocarbon. Now, more than a decade later in 1982, I was back at the INS and disappointed to find my ideas did not fit their expected research goals. And as far as many of the staff at the INS were concerned, I was the same surfing kid from Taranaki who for some crazy reason had 'buggered off to Germany'. Yes, I was still that surfing kid, but I felt completely different.

In the late 1970s, a group of nuclear physicists demonstrated that it was possible to use a particle accelerator to determine the amount of radiocarbon in small samples of carbon. Until then carbon-14 had been measured using Geiger counters, relying on the fact that carbon-14 was radioactive. Geiger counters were a reliable technique, but one that was relatively insensitive and required large carbon samples. The technique could produce a very accurate date for the age of a Stradivarius violin, for example, but imagine telling the owner that you would have to burn the violin to produce enough carbon to date it!

The measurement principle of the particle accelerator technique, which has come to be known as Accelerator Mass Spectrometry (AMS), is completely different. Each of the three carbon isotopes – stable carbon-12 and carbon-13 and radioactive carbon-14 – has a different mass which is easily distinguished using AMS. Rather than waiting for enough carbon-14 to decay to be able to make a measurement with a Geiger counter, AMS makes the measurement immediately by determining the relative amounts of each carbon isotope in a carbon sample. Instead of having to burn a Stradivarius violin to get enough sample to radiocarbon date it, the owner would only need to part with a small speck of wood, a couple of milligrams at most. It was clear that AMS had the potential to revolutionise radiocarbon dating in archaeology.

In keeping with its leading international role in radiocarbon research, the INS had purchased a large secondhand particle accelerator in 1981 and, when I arrived back from Germany, were putting together a science team to develop radiocarbon dating by AMS in New Zealand. The technique was novel, but there was a major snag – worldwide, there were only a handful of AMS laboratories, and reliable production and handling of the tiny carbon samples needed had so far been problematic and elusive.

On my second day back at INS, still moping about my conversation with the director, two nuclear physicists, Rodger Sparks and Gavin Wallace, came to see me. I knew them both from when I worked at the INS before leaving for Germany. Gavin, a friendly Scot, usually wore a white lab coat over shorts and long socks, and was often smoking a pipe with tobacco that was rumoured to be the worst on the planet. Rodger, a New Zealander, had done his PhD in nuclear physics at Utrecht University, in the Netherlands. He and I had a lot in common; he was one of the few people I knew at the INS who could empathise with what I was going through on my return. Now, both of them were looking at me intently.

'We've heard from Bernie that you've a really good background in atmospheric chemistry. I don't suppose you know anything about gas phase chemistry?' asked Rodger.

I said I did – it was one of the things I'd had to learn to be able to measure tiny concentrations of trace gases in the atmosphere. I wondered where this was going and what it had to do with me.

'Well, we need someone to figure out how to make tiny graphite targets out of archaeological samples so that we can use AMS to measure them,' said Rodger. 'Do you think you could turn old bones and shells into graphite for the accelerator ion source?'

'Bones! You did say turn *bones* into graphite?'

When I told Irena what had happened, she thought it was hilarious. But I was totally depressed.

'I've come back to New Zealand with a top PhD in atmospheric chemistry and my job offer is to turn old bones into graphite!' I said, shambling around our tiny living room gulping a horrible beer.

'So, you'll need a lot of bone to make the graphite then?' asked Irena.

'No, this new technique needs only tiny carbon samples, just milligrams,' I said. 'Because it uses mass to measure carbon-14, the sample sizes can be 10,000 times smaller than the existing method, it's almost as if . . .' Suddenly I had an idea.

I remembered one of Dieter Ehhalt's PhD students, Andreas Volz, who had tried to make measurements of carbon-14 in atmospheric carbon monoxide as a way of inferring very small OH concentrations in the atmosphere. In order to get enough carbon-14 for the measurements, he had to strip carbon monoxide from 200 cubic metres of air, a process that took more than three weeks. After he'd collected each sample it took him another four days to measure it. I remembered him swearing about how laborious and error prone the technique was, and if a sample was compromised then weeks of work could be lost. My PhD project with atmospheric formaldehyde had been tough, but he'd almost lost his soul working on his.

Now I was thinking about AMS and the fact that sample sizes could be 10,000 times smaller. Because carbon sample sizes were so tiny, I could see that using AMS had the potential to revolutionise studies of carbon containing trace gases like carbon monoxide, methane and CO_2. I could think of a whole lot of possible applications, including looking at OH concentrations critical to the cleansing properties of the atmosphere, as well as tracing sources of atmospheric methane.

The sources of atmospheric methane were not well understood at the time, and included a mix of biological sources like 'burps' from ruminant animals, wetlands, and fossil sources like gas wells. In the early 1980s, New Zealand had something like seventy million sheep and several million cattle; it seemed likely that the country had relatively high methane emissions from these sources, but no measurements had been made. The next day, when I told Martin Manning my idea, he was really excited. 'Go for it, Dave!' he said.

We could see that if we could measure carbon-14 in atmospheric methane using the new AMS technique then we would be able to provide a world-first estimate of the ratio of biological to fossil carbon in the gas. My ideas seemed promising, but neither of us could have anticipated then just how groundbreaking this research was to be for us both, and for the future of atmospheric chemistry in New Zealand and internationally.

When I next saw Rodger, I told him to sign me up.

'Great,' he said, peering at me through his horn-rimmed glasses. 'So, you've figured out how to turn old bones into graphite then?'

'Nope – not a clue at this stage,' I said.

As usual, when starting a new research project, and with the aid of librarians, I had done an exhaustive literature search. Turning old bones into graphite would have to be one of the most arcane tasks ever, and it was no surprise that the search term 'convert bones into graphite' turned up 'no documents found'. After a fruitless week in several libraries I decided to start with what was known about graphite, from its variety of uses in pencils through to electronics, including electrodes, batteries and solar panels.

There are various grades of graphite, some of which are highly crystalline. With Gavin, I tested a spectroscopic 'off the shelf' grade that I bought from a chemical company. It worked really well in the AMS ion source, producing intense carbon beams that could easily be measured for carbon-14. But how was I going to turn old bones into something like this? For six months I tried every chemical trick I could find to turn the carbon in bones into graphite, but the results were always the same – I'd proudly hand Gavin my latest creation, usually a dubious-looking blackish-grey powder; we'd test it in the accelerator ion source and Gavin would mutter, 'Bloody useless crap, Dave! Where'd you get this?'

After six months of failure I decided to take a step back. A lot of the samples that needed to be radiocarbon dated were actually wood and charcoal, so I decided to have a go at turning those into graphite. Rodger approved of this but reminded me that our real goal was to do bones. 'Yeah, yeah,' I muttered, and got to work.

During the next couple of months, I developed a high temperature vacuum furnace capable of reaching temperatures well above 2000°C. To insulate the system and contain carbon samples, I used boron nitride, a ceramic originally designed for shielding spacecraft on re-entry to Earth's atmosphere. After several failures, I tried using sawdust from rimu wood. Soon I made a highly crystalline black powder. I raced down to see Gavin.

'You got some more carbon crap for me to test then?' he asked.

Trembling with anticipation, I watched as he mounted the powder in the ion source and turned on the machine. Both of us watched in disbelief as the sample produced a large stable carbon beam, easily measurable for carbon-14 and exactly what was needed for radiocarbon dating using AMS.

'Congratulations, Dave – bloody good!' said Gavin. 'But where are the old bone samples?'

'Bugger off!' I shouted.

Settling in had been difficult at first, but we threw ourselves into life in New Zealand and soon adapted. Both Irena and I enjoyed the outdoors, and it was great exploring the back country with Greg and Suzanne. Our biggest problem was the house, which was damp and cold in winter and cramped for the four of us. We were expecting another child at the end of 1983 and it was clear we were either going to have to extend the house or find another one.

After a lot of thought we decided to extend the existing house, taking out a large mortgage to pay for the work. The few months that it took to complete the work were chaotic. Unfortunately, we had chosen a less than reputable building firm who ripped us off with substandard work. Compounding this was Skip, arriving from California with a new German girlfriend and a bad attitude, freeloading and putting extra emotional stress on me and Irena. This was the last time I saw him. Years later, Skip was tragically killed by a rattlesnake in the California desert. I'd really liked him and this was a shock.

When I think back, it was sheer hell; an El Niño year with

constant gale-force winds and huge scrub fires on the gorse-covered hills behind us, months of fruitless work trying to make graphite, living with the horrible building project which seemed to go on and on, and missing Germany all the while. On my scientist's salary, with crippling mortgage repayments, we were having difficulty making ends meet. At one stage we didn't even have enough money to buy Irena new shoes.

But everything changed in November with the arrival of our second daughter, Johanna, at Lower Hutt Hospital. What a bundle of joy, transforming our lives and completing our family. And during this time, I'd figured out how to make the first successful graphite targets for AMS carbon-14 dating at INS. Yes, we were poor financially, but that summer as we rolled into 1984, the newly extended and insulated house was filled with laughter.

A mix of coincidence, hard work and maybe 1% inspiration have played a major role in the scientific developments I've made. I don't know why, but for some reason in early 1984 I'd been looking at work involving the iron and steel industry, when I noticed that researchers often found small deposits of graphite formed in iron after iron ore was reduced during the smelting process. I was intrigued and immediately began some experiments using iron powder, CO_2 and hydrogen. The reactions were similar to those used in World War Two to make gas to power vehicles when petrol and diesel was in short supply. After experimenting with the reaction at different temperatures and using pressure detection techniques from my electronics background, I soon found that I could quantitatively and reliably produce graphite from tiny samples of CO_2. As a final step, I developed the method so the graphite formed directly in a copper target holder, which could be mounted in the accelerator ion source. This avoided the almost impossible chore of trying to manually transfer a milligram of graphite into a target holder.

This was the breakthrough that I had been looking for. But what did it have to do with turning bones into graphite? Well, any carbon sample to be radiocarbon dated can be turned into CO_2 with suitable pre-treatment. The pre-treatment needed to extract carbon from atmospheric methane would be very different to the method required for bones, but the end result was the same. Both kinds of samples could be turned into CO_2, after which the process I'd just developed would produce graphite ready for carbon-14 measurement on the accelerator.

The method was quick and versatile, and I lost no time scaling up the process so that we could make multiple graphite targets in a few hours. Radiocarbon dating using AMS was up and running at INS. Soon we began getting orders from all over the world, to date tiny samples of carbon.

A smiling Gavin congratulated me as Rodger poured sparkling wine into a filthy beaker in the corner of the new lab I'd set up for the graphite target preparation. But the celebrations were called short by an urgent job: a curator from a museum in Melbourne wanted us to verify the age of one of their Egyptian mummies. They were worried that it might be a fake, and had sent us a tiny finger bone as a sample for immediate analysis.

That night I worked late in the lab with the mummy sample. As I converted part of the finger bone to CO_2 and transferred this to a small graphitisation furnace, I thought about the fact that this sample had come from what used to be a human being. I was all alone in the institute and it was dark. I'm not superstitious, but I shivered involuntarily. Just then I heard a noise outside the lab – with the recent Indiana Jones movie fresh in my mind, I slowly opened the lab door and peered into the corridor. I was sure there was something there but couldn't see anything. As soon as I'd finished making the graphite target, I raced home.

The next day, Gavin and Rodger measured the graphite that I'd prepared from the mummy bone and came up with a date of about 5000 years. This was in line with what the museum had expected and a big relief for them. Our reputation as a reliable supplier of AMS radiocarbon dates was growing. Over the next couple of decades, what an atmospheric chemist and a couple of nuclear physicists had started turned into a million-dollar business.

Martin Manning and I were becoming more and more concerned about increasing levels of atmospheric CO_2 and methane, and we continued to speak out. At a scientific conference in Wellington I spoke about atmospheric methane, including the large quantities of the gas burped out by ruminants like sheep. A journalist at the meeting picked up on my comments and soon they were reported in newspapers around the world, with headlines like 'Sheep Gas Blowing in the Wind'. Even more bizarrely, my comments were turned into a question on a Trivial Pursuit card: 'Which animal did New Zealand geophysicist Dr David Lowe say was adding to global warming?' The answer: 'A sheep!'

By the mid-1980s more and more research was documenting evidence of atmospheric degradation, including stratospheric ozone depletion and widespread dispersal of both gaseous and particulate pollution from cities. Both the US and the Soviet Union bristled with nuclear weapons and potential catastrophic 'nuclear winter' damage to the atmosphere in the event of a war. I wrote several articles on the subject, including a paper for *The New Zealand Journal of Science*,[12] and was invited by Helen Clark to address the government on the issue at a parliamentary select committee. It was a time of political and economic turmoil in New Zealand, marking the end of Robert Muldoon's disastrous term and the beginning of David Lange's radical economic and political policies like the banning of nuclear-armed and powered ships from entering New Zealand harbours. This passed into New Zealand law in 1987, an action which angered the US and other 'friendly' nuclear-armed countries. Tensions were already high due to the sinking of the *Rainbow Warrior* in 1985, a Greenpeace vessel destined to protest at Mururoa Atoll in French Polynesia, a hugely unpopular French nuclear weapons test site. The French government had sanctioned the bombing of the *Rainbow Warrior*, and although the wider nuclear energy issue was not directly related, these events led to a growing awareness among New Zealanders of the dangers of increasing levels of atmospheric CO_2 and methane. I began speaking on these issues at public groups like local tramping, Lions and Rotary clubs. At the time I did not meet many climate deniers in the audiences, but oddly the few that I did meet seemed to be both pro-nuclear power and nuclear weapons. I remember a furious argument with one individual who exclaimed, 'How dare New Zealand not want to be protected by the nuclear umbrella!'

Now that the graphite system was working at INS, my research focus turned to how to get enough methane from an air sample at Baring Head to make a carbon-14 measurement. This was a huge ask, but I had already discovered that it was technically feasible due to the tiny sample capability of the accelerator system; we had shown that we only needed a milligram of carbon in the form of graphite to make the carbon-14 measurement. But how could I extract the tiny quantity of methane in air, only 1.6 ppm, remove the carbon from it, and turn that into a minuscule graphite sample for the accelerator? A simple calculation showed that I needed to process at least a cubic metre of air at Baring Head to provide enough carbon derived from atmospheric methane to make the carbon-14 measurement on the accelerator. As had happened several times before in my career, a possible answer came to me from another area of science. I knew that activated charcoal at cryogenic temperatures could trap methane. If I sucked a cubic metre of air at Baring Head through a series of activated charcoal traps immersed in liquid nitrogen, I should be able to trap the tiny amount of atmospheric methane in the air. The first runs were very promising, and I was able to extract the methane afterwards at INS, converting the carbon in it to graphite for the accelerator. Martin Manning and I celebrated the encouraging initial carbon-14 results.

But the technique was extremely dangerous because, not only did the activated charcoal trap atmospheric methane, it also trapped liquid oxygen – producing a potentially explosive mixture. Mixtures of liquid oxygen and activated charcoal were widely used after World

202

War Two in Germany to remove the remains of demolished buildings wrecked by allied bombing. I'd figured out how to avoid this disastrous situation using precise pressure control. But this required careful attention to detail – operator mistakes were likely to be very dangerous. The collection system worked well and it was exciting to see the first carbon-14 methane measurements on the AMS system at INS. We needed to increase the amount of data so I trained a science technician, Greg Drummond, to help me with the sample collection out at Baring Head. At first things went well, but after a couple of weeks I had an ominous phone call from Steve O'Neill, the lighthouse keeper at Baring Head.

The inevitable had happened. The technique was difficult to run. Greg had made a mistake with a valve closure and one of the liquid nitrogen traps had built up pressure and exploded. After that, he'd shut everything down and stumbled over to the lighthouse keeper's house, where Steve had settled his nerves with whisky. When Greg arrived back at the INS lab, he seemed oddly cheerful but reeked of whisky. I gave him the rest of the day off.

I abandoned the technique as being too dangerous and switched to a method using large air samples pumped into 70-litre stainless-steel tanks using a clean air compressor. The tanks had been originally designed for use as LPG cylinders in trucks but, after modification, were ideal for the atmospheric methane research.

Over the next six months we collected about twenty large air samples at Baring Head and processed them for atmospheric methane on a vacuum extraction line that I'd designed in the lab at INS. Once the methane was separated from the air sample, it could be converted into graphite using the same system that I'd used for the Egyptian mummy bone. The graphite targets were then measured on the AMS system to provide the first ever carbon-14 in atmospheric methane data. Working as a team, we used the data and a mathematical model developed by Martin Manning to show that about a third of the methane in the atmosphere must have been derived from fossil sources. It was a sensational result, a world first scientific breakthrough, one of the highlights of my career and published in the prestigious scientific journal *Nature*.[13] The results were widely reported in the media and we were congratulated by our peers around the world. We had shown that the AMS carbon-14 technique could lead to unexpected discoveries in atmospheric chemistry, with important research and policy consequences. Subsequent work would reveal that a lot of the fossil methane in the atmosphere was probably coming from sources like leaks in gas pipelines in the northern hemisphere. Since our initial discovery, carbon-14 measurements by AMS have become widely used in atmospheric chemistry, with some groups even able to detect carbon-14 in the vanishingly small amounts of methane extracted from air tens of thousands of years old, trapped in bubbles in polar ice cores.

We'd demonstrated the power of isotopic measurements to find out things about the atmosphere that couldn't be determined in any other way. Using AMS and carbon-14 to understand the sources of atmospheric methane is a little like using DNA techniques to trace humans – we'd used what was effectively an isotopic fingerprint to establish the sources of this climatically very important gas.

I had been bitterly disappointed when I'd arrived back in New Zealand from Germany and had to work on the AMS carbon-14 technique, something I'd considered completely remote from atmospheric research. But the atmospheric chemistry I'd learned in Germany complemented the seemingly unrelated isotopic techniques at the INS and helped us launch what has become an internationally recognised and productive atmospheric chemistry research group in New Zealand. Now, decades later, I often reflect on what that single task of turning mere bones into graphite led to.

CHAPTER 13

ROCKY MOUNTAIN HIGH, COLORADO

In the 1980s, the concentration of atmospheric methane was increasing at a rate approaching 1% per year. Our work with carbon-14 in methane at Baring Head had shown that about a third of atmospheric methane's sources were fossil in origin. But this implied that the remaining sources were probably biological. What were they? If a major part turned out to be methane burps from ruminant animals, for example, then New Zealand – with its massive numbers of sheep and cattle – could be a relatively large source of atmospheric methane, with implications for climate change. It was essential that we expand our research to find answers.

In early 1988 I was contacted by Ralph Cicerone, a director at the US National Center for Atmospheric Research (NCAR) based in Boulder, Colorado. Ralph was a colleague of Dieter Ehhalt's whom I'd met at a couple of international meetings as well as at Scripps. He agreed with Martin Manning and me that carbon-14 would be a very important tool in atmospheric chemistry and invited me to spend a year working with his group at NCAR who were carrying out a lot of research on biological sources of methane including rice paddies, marshes, landfills and ruminant animals.

Boulder itself was becoming a centre of excellence in atmospheric science. Not only was NCAR based there but also NOAA, a large US government lab involved in meteorology, stratospheric ozone research and systematic atmospheric monitoring for CO_2 and methane. A couple of the PhD students I'd studied with in Germany worked for NOAA, as well as a close personal friend of mine from Jülich.

When I talked the invitation over with Martin Manning and others at INS, it was obvious a year at NCAR and in the Boulder atmospheric science community would be of huge benefit to our own research goals. Although we had a lot of expertise in isotopic techniques at INS, we had limited experience with gas chromatography, which would be an essential tool if we wanted to track increases in gases like atmospheric methane and carbon monoxide. But there were a couple of snags. The main one was my personal situation with Irena and the children; there was no way I would leave them for a year! Thankfully, Ralph was willing to extend the invite to include my family. NCAR promised to take care of all the details – flights, schools and an apartment.

The remaining snag was that carbon-14 dating at the INS using AMS was still totally reliant on the skills and techniques I'd built up making graphite targets. At first, I'd worked very long hours running the graphite lab. The workload was shared by Greg Drummond, whom we'd nearly blown up the year before at Baring Head. Both Rodger and Gavin were adamant that dropping my role in the lab to move to Colorado for a year would be a big setback for the carbon-14 AMS programme. Then, out of the blue, I received an enquiry from an American atmospheric chemist with an unusual name, Ed Dlugokencky. Ed was working in Boulder at NOAA and had heard I was thinking about moving there for a year. He wondered whether he could work in my position running the graphite lab and atmospheric science projects for a year while I was away. He was a keen cyclist and wanted to experience living in New Zealand and working at INS. It was an ideal solution and would kick-start his career as the foremost world expert on increases in atmospheric methane.

It was the summer of 1988 and we'd been living in Boulder for about two months. The transition for our family from New Zealand had been remarkably painless thanks to the support we'd had from NCAR and our friends who lived in Boulder.

Boulder is an inspiring place to live, with the natural environment dominated by mountains rising abruptly from a flat plain that stretches almost 1000 kilometres to the east. It features the world famous University of Colorado, several labs specialising in atmospheric science, and is a magnet for academics from all over the US as well as overseas. The city is well served with bike lanes and paths and surrounded by a 'green belt' protecting the residential core from new development. The weather sees more than 300 fine days a year, hot dry summers and cold snowy winters. It's a paradise for anyone who is into outdoor activities like skiing in the winter or hiking and biking in the summer. As a keen cyclist, I was soon biking to work at NCAR every day, joining hundreds of others in the most cycle-friendly city in the US. Greg and Suzanne had started at a local school, Bear Creek Elementary, only five minutes' walk from our apartment, and Johanna attended a Montessori preschool also close by.

On my first day at NCAR I was introduced to Stan Tyler, with whom I would work for the next year. Stan is one of the most colourful and fun characters I have ever worked with. From day one it was jokes and laughter as we forged a solid and productive working relationship. Stan had a PhD in nuclear physics and struck me as one of the most intelligent people I knew. He had an incredible memory and attention to detail, but is also one of the most impractical people I've seen in a lab. He'd been hired by Ralph Cicerone to work with the NCAR group on stable isotopes in methane sources like rice paddies and wetlands.

Before leaving New Zealand, I had established that NCAR was buying a state-of-the-art stable isotope ratio mass spectrometer for Stan so that we could carry out the work we had planned together. At the INS I'd worked with accelerator mass spectrometry to determine levels of carbon-14 in atmospheric methane. But to really understand different biological sources of atmospheric methane, it was going to be essential to make high-precision measurements of the stable isotopes in methane such as carbon-13. These measurements are made using the type of equipment NCAR had ordered, simply called 'mass specs'.

Shortly after my arrival in Boulder, the German-made mass spec was delivered and due to be installed by specialist engineers. In Germany all equipment like this is powered by 230 volts AC, which is also the standard voltage used in New Zealand. But the standard voltage in the US is only 110 volts AC. The German company supplying the mass spec was prepared to provide an inverter to convert the voltage, but for a hefty price. To save money, Stan and the NCAR electrician, Jeb, had decided to get one made locally. Jeb had a contact at a local prison with a workshop making electrical equipment like inverters. Stan contracted the business to build the inverter and it was delivered with the mass spec.

On the day the mass spec was to be installed in Stan's lab, I watched as Jeb wired up the plugs on the inverter. The gear looked homemade and distinctly suspect compared to the sophisticated mass spec that it was supposed to power.

'Have you checked the output voltage of the inverter?' I asked.

'Nope – but I don't need to. They will have checked out all that stuff at the prison,' he said, as he plugged the mass spec into the inverter. There was a tremendous flash, a couple of bangs, and Jeb's head was enveloped in smoke. He staggered back from the equipment and sat down, pale-faced, while Stan called for a paramedic.

The mass spec had been badly damaged. When I looked inside its control computer, every single integrated circuit on its motherboard had a crater melted into it. Jeb came to see me the next day and took me aside, ashen faced.

'After I got back home from them medics, I found my digital watch had reset itself, and I couldn't hear so well. There's worse too. Last night I couldn't . . . '

'Yeah yeah, okay, Jeb. Too much information!'

Ralph Cicerone came to see me and apologised for the debacle. NCAR would do their best to find replacement parts, but repairing the mass spec could take up to a month. I assured him it wouldn't be a problem and resolved to use the time to learn more about gas chromatography.

*

Winter arrived with a rush of snow. One day it was fine and sunny, and the next morning we woke to find half a metre of snow outside. The children lost no time running out to make snow angels, falling on their backs in the snow with their hands outstretched. We went skiing in the Rockies, and on our first trip Stan asked if he could come with us to the world-famous Copper Mountain. We readily agreed – he'd told us about the 'black' slopes he had skied on which were reserved for experts. Greg was eight and in awe of Stan, an American who skied on black slopes – wow! As a family we had so far stuck to the safety of the 'green' beginner or 'blue' intermediate slopes. A family service on the mountain provided babysitting, lunch and children's ski lessons for Suzanne and Johanna, leaving Greg, Irena and I free to explore the whole of Copper Mountain with Stan.

Our first run with Stan was unbelievable. A huge high-speed chairlift took the four of us well up the mountain to a spot where the maps showed a starting point for a variety of black, blue and green trails.

'Hey Stan, there are the black trails over there,' I said. 'You take one of those and we'll head off down one of these intermediate runs. We can meet you at the main café at lunchtime?'

'Tell you what,' he said. 'I'll come down with you for starters.' And off we went, weaving our way down the slope on a broad swathe of snow between lines of trees.

'Where's Stan?' called Greg, as the three of us came to a stop near a snow-covered bank.

'We thought he was with you,' said Irena. We looked up to see Stan careering wildly across the slope above us. With his hands flailing, he skied straight into a tree. I'll never forget the expressions that crossed Greg's face when he saw Stan with his arms wrapped round that tree. We laughed as we trudged up the slope to rescue him, and Stan laughed with us.

*

After a month, the mass spec was repaired and we began measuring carbon-13 and carbon-14 in atmospheric methane from air samples taken on a high ridge in the Rockies. To do this, I designed a vacuum extraction line which allowed us to make high-precision carbon-13 measurements on the mass spec and carbon-14 samples which we sent back to New Zealand to be measured on the accelerator back at INS.

Ralph Cicerone asked Stan and I to provide measurements for an experiment aimed at tracking a shift in carbon isotopes when methane is removed from the atmosphere by the OH radical. This followed original work done by Dieter Ehhalt in Jülich, and required long and careful hours in Stan's lab. The effort paid off. Our results, published in the *Journal of Geophysical Research*, have since been cited hundreds of times and are widely used in modelling studies estimating atmospheric methane sources and sinks,[14] essential for determining the atmospheric budget of methane.

NCAR is on a steep-edged, flat-topped mesa more than 200 vertical metres higher than Boulder, which is already 1600 metres above sea level. Being high in the foothills of the Rockies means it can snow in Boulder without warning at any time of year, even summer. In the winter snowfalls are heavy and the road to the NCAR was often blocked. Stan's lab was in a basement with no windows, and we were usually unaware of conditions outside.

One day we'd been working through the afternoon when we noticed that the NCAR building seemed deserted. Security had told the research staff about a heavy snow warning and everyone had been sent home early, but they'd forgotten to tell me and Stan. When we looked outside it was a white world with just a couple of snowy humps that marked the few remaining cars in the carpark. One of these was Stan's. As usual I'd ridden to work that morning on a bike, and it was clear that I would not be going anywhere with that. We managed to dig Stan's car clear of snow, but it refused to start. Nearby there was a Jeep with its motor running, and I asked the driver if he could help us with a jump start. The man apologised that he didn't have any jumper cables.

'No problem,' said Stan, and fetched a heavy black wire and a thin piece of red hook-up wire with tiny alligator clips.

'Stan, this is not going to work,' I said.

'It'll work!' assured Stan. 'You see, the thick black wire takes all the current.' The jeep owner's face was a study in disbelief. Even my son Greg at the age of eight knew what an electrical circuit was – two equal wires were required to complete a circuit; if one was thin like this red wire, the outcome would be dodgy.

Stan clambered into his old Datsun. 'You connect it up and I'll start the engine,' he said. The result was spectacular. As soon as Stan tried to crank the Datsun's engine, the thin red wire burst into flame.

'If you dudes don't mind, I'm out of here,' said the Jeep owner. He drove away, leaving me and Stan alone in the snow.

After a frantic call for help, Irena somehow managed to drive up to the NCAR mesa to pick us up in our old Plymouth Reliant station wagon, and got us back to our apartment. Soon after, all of the roads in Boulder were blocked with snow, and Stan – who lived another 10 kilometres away – had to spend a couple of nights with us. The winter storm was one of the coldest in a decade, with the temperature dropping to -35°C overnight. The next morning, I showed the children how you could take a cup of water from the kitchen, open the front door and fling the water into polar air. Before the water could fall it froze into ice, which tinkled as it hit the ground.

The year spent living and working in Colorado was superb in many ways. The experience of going to a school in another country has stayed with our children and is part of who they are – living in the Rocky Mountains environment of Boulder was a privilege. And for me, scientifically, it was a success. I achieved a lot through my research at NCAR with Stan and had made valuable contacts at both NOAA and the University of Colorado for our New Zealand programme. In Boulder at the atmospheric science institutes, I could share my drive and resolve for the climate with likeminded people. There was a joint concern about the rapidly changing composition of the atmosphere, that this was a global issue needing an international response.

Ed Dlugokencky, who worked at the INS during my absence,

returned to Boulder, first working with Stan at NCAR and then with NOAA. During the next thirty years he's had a brilliant career measuring and documenting the unrelenting rise of various atmospheric trace gases measured at a world network of sites he helped set up. His work is widely cited and forms the basis for understanding increases in atmospheric methane. I've kept in contact with him all that time and he is one of our closest friends – after all, his work at the INS has made him part-Kiwi and he understands New Zealanders. And for my part, I am proud to have lived and worked in Colorado, learning from talented and generous scientists. Almost without exception the people we met in Boulder take care of their precious environment and are aware of the fragility of the atmosphere.

CHAPTER 16

NIWA AND GRETA POINT

On my return to the INS from NCAR at the end of 1989 I found Gavin and Rodger waiting for me. The institute had been overwhelmed by requests from all over the world for carbon-14 dates using the accelerator, and there was a backlog. Furthermore, our atmospheric chemistry work was gaining international recognition. In my absence Martin had been successful promoting the programme in New Zealand and had attracted a lot of new government funding. Over the next couple of years this allowed us to expand from two scientists to five, spending most of their time working on projects related to atmospheric chemistry. This included Keith Lassey, another nuclear physicist, who has become one of New Zealand's top experts on methane emissions from ruminant animals; and Gordon Brailsford, an outstanding experimentalist responsible for the increasing success of Baring Head with its large number of atmospheric measurement projects. Before I left for NCAR we'd been joined by Carl Brenninkmeijer from the Netherlands. An expert with mass spectrometry, he had an appreciation for the role of stable isotopes and carbon-14 in atmospheric chemistry. Carl developed a brilliant technique for measuring carbon-14 in atmospheric carbon monoxide, which received a lot of attention when it was published in the prestigious scientific journal *Nature*.[15] Thirty years later, this and other techniques we developed are still being used in New Zealand and internationally.

The early 1990s were among the most productive in my career. There was a constant feeling of excitement within the atmospheric science group (as we came to be known), with an increasing number of new measurements and discoveries reported at meetings and published in international scientific journals. We attracted many visiting scientists from overseas as well as PhD students, all of whom contributed to the group. We became an integral part of international scientific collaboration focused on understanding the implications of increased human damage to the atmosphere. Our measurements documented increases in atmospheric methane and CO_2, and our carbon isotopic research helped infer the causes of the changes.

There was increasing push-back at international climate change research, including my own, from fossil fuel companies and other organisations. By the 1990s, oil was a trillion-dollar industry driven by profits and a growth model that many economists warned was incompatible with life on a finite planet. This was classic market failure – unchecked growth leading to irreparable damage of the very ecosystems that supported human life and the world economy in the first place. The Marshall Institute, a conservative thinktank based in the US, began promoting fringe views on climate change science and received extensive financial support from the oil companies. The institute ran what it termed 'a critical examination of the scientific basis for global climate change policy', later recognised as a 'central cog in the denial machine'.[16] They had a large impact politically in the US, including on the George HW Bush administration, where the president pledged to 'meet the Greenhouse effect with the Whitehouse effect'. Later, Naomi Oreskes and Erik M. Conway, in their very effective book *Merchants of Doubt: How a handful of scientists obscured the truth on issues from tobacco smoke to global warming,*[17] identified a few contrarian scientists associated with conservative thinktanks like the Marshall Institute who fought the scientific consensus and spread misinformation, confusion and doubt about global warming and the dangers of tobacco smoke inhalation. For many years I'd been giving talks on atmospheric greenhouse gases and climate change to community and other groups, including high schools and universities. But the 1990s marked a change, as I began to battle climate sceptics –

now called climate deniers, many of them inspired by the activities of organisations like the Marshall Institute. Some of the more notorious climate change sceptics in New Zealand were associated with the coal industry and used pseudoscience arguments. A classic example would be, 'The climate is changing but it's driven by changes in the sun.' Yes, the sun's output is changing, but that's insignificant compared to the changes to the atmosphere caused by humans. By the time I'd explained this at one of my community climate change talks, I would have lost my audience's interest – which was exactly the intent of the sceptic. Newspapers would publish 'false balance' articles giving equal weighting to the science delivered by thousands of climate scientists and the pseudoscience by one climate sceptic. The public was becoming confused about the reality of climate change, and we even faced arguments from our own science administrators who had backgrounds in accountancy and management, and had little or no scientific education. I'd been through many ups and downs during my career, but the frustration of having to deal with deniers peddling scientific mistruths was undermining and demoralising. The deniers were so successful that there were times when I even faced uphill battles talking about climate change to my own friends and family.

In 1992, suddenly and without warning, Mum died. Her health had been declining for years, but she had been a rock and a link to my childhood and father. Her belief in me had carried me through countless trials and dark places. She had celebrated my successes and was proud of what I had achieved in the world of science. Her passing left a huge void.

In the 1990s, bureaucrats and politicians in New Zealand were busy dreaming up new ways of running education, health, science and other government ministries. One casualty was the DSIR, the premier government department and primary provider of scientific research in New Zealand which underpinned vital achievements and contributions to New Zealand's environment and economy. Athol Rafter's carbon-14 work had thrived under DSIR management and his institute, the INS, had become world-famous. In mid-1992 the DSIR was dissolved. Pieces of it, with elements from other government departments, were spread around ten newly formed Crown Research Institutes (CRIs).

This had a major impact on the atmospheric chemistry group that Martin and I had been building. First of all, there was a requirement that the CRIs be 'financially viable' and operate on commercial lines. But more than this, the CRIs were to be 'discipline-based', a kind of vertical structure which ignored the fact that most of the science in New Zealand, including ours, was multi-disciplinary. Because our group was involved in atmospheric chemistry, we were allocated to the National Institute of Water and Atmospheric Research (NIWA) and split off from the equipment development we had helped pioneer at INS, including the accelerator and mass spec work which now belonged to another CRI called the Geological and Nuclear Sciences (GNS). If I wanted to use the equipment I had built, I had to pay a negotiated fee to GNS. Former workmates, including Rodger and Gavin, became competitors overnight, because they were competing for the same government science funds. This was part of the initial CRI model that we were told would improve the efficiency of science in New Zealand. Within a couple of months, we were shifted from our labs and offices at the INS to sub-standard accommodation in Lower Hutt – a 1950s building which was unbearably hot in summer and bitterly cold in winter. Our colleagues were now meteorologists and upper atmospheric physicists.

The move to the CRIs was a huge impediment to our research. I was grieving the loss of Mum, and to add to my troubles this artificial structure created by bureaucrats seemed set to destroy everything I'd worked for. We tried having atmospheric methane samples run on the mass spectrometer at GNS. The results were a disaster, with weeks of preparation wasted due to contamination by an inexperienced operator. Rather than send more samples there, we started saving them in sealed glass vials in the hope that perhaps we'd find a way of measuring them later on. I was bitterly disappointed and could see no way forward.

One evening when I was riding my bike home after a miserable winter's day in the NIWA building, and shortly after the debacle with the contaminated samples, I was hit by a car. The driver of the car failed to give way, hit me head on and said that I was in her blind spot. An enduring memory I have is lying on the road, hurting everywhere, and seeing my bike corkscrewing through the air after the impact. I was badly bruised and had to have a cracked femur checked out in the local hospital. Between Mum passing away, the CRIs and now this, I felt that things couldn't get much worse.

But after a terrible first year at NIWA, out of the blue one day Martin and I were contacted by Rick Pridmore, an American biologist who worked for NIWA in Hamilton. He was NIWA's new research director and visited me the following day. He'd read all my papers and, understanding the significance of our research, sympathised with how we'd lost our gear in the change to CRIs.

'Do you mind taking me up to GNS so I can have a look?' he asked. We jumped into my old wind surfing van and I drove him to the site of our old equipment, where Martin and I had set up the atmospheric chemistry programme. I was astonished to find that Rick immediately understood everything I showed him and remember thinking what a great choice for a research director he was. After we'd returned to the dreadful old building where NIWA had housed us, Rick lost no time in telling me his plans.

Impressed with our research, he said he wanted me to design a purpose-built lab for NIWA at Greta Point in Wellington. I could buy a couple of new mass specs and other gear I needed to get our group and others in NIWA up to speed. 'How about a million dollars?' he suggested as a budget. This was a huge amount of money, especially in the early 1990s. The next day, Rick sent me an email:

Dave, I mean this, don't compromise on the lab facilities you want. I will support you and your team. You guys do great work and have great spirit. I don't want to change either.
Cheers, Rick

After a dreadful year, here was a person who not only could see the scientific value of what we were doing, but was prepared to back us. When I told the rest of the group about the message, they were blown away – except for Martin, who didn't seem surprised. 'I've always believed in the value of our work,' he said.

With Rick's encouragement I spent a month visiting mass spec users in the US, Germany and the UK. With the exception of two groups – one at Indiana University Bloomington and one including Carl Brenninkmeijer, who had moved from our group at the INS back to Europe – no one in the world had tried to make the kind of atmospheric measurements we were proposing. When I visited the Finnigan MAT mass spec factory in Bremen, Germany, their chief development engineer, Willi Brandt, threw up his hands and said, 'Our machines are designed for making isotope measurements of carbon in rocks, not in tiny amounts of atmospheric trace gases. You'll never get it to work!' I assured him that I would.

A year later we had the lab up and running at NIWA Greta Point, Wellington, with state-of-the-art mass specs designed and bought from the Finnigan MAT factory. By this stage we had hundreds of vials of unmeasured atmospheric samples collected during the eighteen months that we'd been separated from our gear. When we measured them on the mass spec they showed a series of fascinating data from Antarctica and Baring Head.

Our reputation was growing. By the late 1990s we were attracting scientific visitors from all over the world, including Paul Crutzen, who in 1995 was one of three scientists awarded the Nobel Prize for Chemistry for research into the causes of destruction of stratospheric ozone. At any one time we had up to fourteen people working on different atmospheric chemistry projects, and I suggested we officially name our group TROPAC, short for Tropospheric Atmospheric Chemistry.

The new lab facilities and equipment that Rick Pridmore had encouraged me to develop were outstanding and allowed us to tackle a wide variety of research problems, especially those needing a combination of isotopic and conventional measurement techniques. At Baring Head, we were observing rapid changes in atmospheric CO_2 and methane; by the early 2000s the average growth rate of CO_2 was approaching 2 ppm per year, double the rate when I first began measuring it thirty years earlier. By this stage the Kyoto Protocol, the first international agreement aimed at limiting greenhouse gas emissions, had been adopted and, with Rick Pridmore, I was proud to be in attendance at the Parliament Buildings to witness Helen Clark as prime minister ratifying New Zealand's commitment to this important climate agreement. From our work and others, it was clear that a substantial fraction of New Zealand's greenhouse gas emissions was methane from agricultural sources. This became politically important. Keith Lassey from our TROPAC group, along with other scientific institutes like New Zealand's agricultural CRI, helped provide policymakers with a sound scientific basis for decision-making. However, despite the validity of NIWA's early data and projections showing that agricultural emissions were New Zealand's largest source of greenhouse gases, the research was not acted on politically. Primary production including dairy, lamb and beef accounted for over 40% of all exports, and including it in the 2008 emissions trading scheme was put into the politically 'too hard' basket.

Many of our overseas visitors wanted to collaborate with us on isotopic techniques and standards. In 2000, about five years after we had bought the Finnigan MAT mass spec, we had a visit from Willi Brandt, the German engineer who had told me what I proposed was impossible. He had left the factory in Bremen and was running a very successful atmospheric science group at a Max Planck Institute in Jena, formerly in East Germany, using modifications of the very techniques I had spoken to him about. 'Looks like you were right,' he said, shaking my hand. 'I owe you a beer!'

*

To my left I could see the swells lining up as they broke around a headland a few kilometres away. To my right was Kapiti Island, a bird sanctuary off the southern coast of the North Island. My board slapped the waves as I leaned back in my harness, revelling in the steady wind powering me over the swells. Ten years earlier, as a substitute for surfing, I had taken up windsurfing, and the sport had enthralled me and my son Greg with its endless conditions and challenges. It combined my love of the wind and the waves.

In the late 1990s our family had moved to a large hillside property along the coast in Plimmerton, about 30 kilometres north of Wellington. The land had been severely compromised by about fifty non-native macrocarpa trees, with weeds and creepers growing unchecked and smothering the few native plants still left. Irena and I had resolved to 'heal the land' by rewilding it with native species, but the first job was to remove the macrocarpas. Full of confidence I bought a top of the line German chainsaw and read in the owner's manual how to fell a tree. *Cut a felling notch in the direction you want the tree to fall and pick two escape routes on the other side of the tree. When the tree begins to fall, switch the saw off and walk away from the tree along an escape route.* On my first attempt everything went by the book until the tree began to fall – when it did, my bloodstream immediately spiked adrenaline. I ran away from the tree carrying the still-running chainsaw, knowing that this activity was not for me! It was in Plimmerton that our children became adults. They loved the natural environment and appreciated the conservation and restoration projects Irena and I were involved in. I'd already spent years windsurfing with Greg, and as he became a young man he developed the same feelings for the wind and waves that I did. He is now just as interested in climate science, every couple of weeks sending me news on renewable energy and carbon emissions reductions. He spends his winters in the South Island ski-touring in remote alpine environments and summers windsurfing around Wellington. Suzanne completed an environmental science degree at the University of Waikato and has a successful career in contaminated land restoration and wastewater management. Johanna is a wonderful artist and photographer inspired by the environment around her. We spent eighteen years at Plimmerton planting more than 2000 native trees and shrubs, trapping pests and setting bait for rats. It's a source of pride that today the hillside carries a thick stand of native bush with a year-round supply of fruits and flowers, supporting many species of native birds. It's a legacy that will outlive us.

In the early 2000s NIWA prospered under Rick Pridmore, and the TROPAC group continued to expand our work with more visitors and PhD students. One such visitor was Australian David Etheridge, a world expert on ice cores from polar regions. Ice in Antarctica and the Arctic contains bubbles of air trapped when snow compacts to form solid ice. Because deep ice cores can be several hundred thousand years old, studying the trapped air can provide information on past climate change events including the transition out of ice ages. Ice ages can be viewed as a kind of climate change caused by slowly changing natural events over very long time scales, meaning that data from ice cores is highly relevant to understanding current rapid climate change driven by humans. David Etheridge invited us to collaborate with him, analysing methane in air extracted from bubbles in ice cores drilled from Law Dome glacier in Antarctica.

This marked the beginning of a new phase of research for TROPAC. I supervised a talented PhD student, Dom Ferretti, with the development of an innovative mass spec technique to measure the isotopes in the CO_2 and methane contained in tiny air samples like the ones from ice cores. The technique is now used in many labs around the world. Dom subsequently worked with groups in the US and Australia to find unexpected levels of methane in the atmosphere during the last 2000 years. The results were published in the leading American journal *Science* and suggested that humans, through widespread burning of forests and grasses, were already having a significant impact on the atmosphere more than 1500 years ago.[18]

By the early 2000s I had helped supervise a number of PhD students, something I found rewarding and considered very important. Rather than working on my own, I was part of a dynamic team of highly talented researchers making significant discoveries about the changing atmosphere. I decided that it was a good time for me to step back from frontline research. It was clear there needed to be growing public awareness of climate change issues, and more young scientists would be needed to research the changing atmosphere and environment.

I began teaching a fourth-year honours course in atmospheric chemistry at Victoria University of Wellington. I've always enjoyed teaching, but I was unprepared for the amount of work involved. Although I was on top of the science, preparing the course took weeks of work, most of which was in my own time. Because NIWA required cost recovery on most activities and the time involved was difficult to justify, my time was 'donated'. I taught the students basic atmospheric chemistry and physics, and introduced them to the fundamental processes controlling chemical cycles in the atmosphere and how these have been dramatically changed by human industry and agriculture. At this point I had been living with the horror of climate change for over thirty years, and had faced down sceptics, deniers and foot-dragging bureaucrats. But when I explained the consequences of unchecked atmospheric emissions to the students in my classes, I was unprepared for their reaction. I saw in them what I'd felt during my whole working life: raw anger, they were seething with the unfairness of it. Younger generations have the most to lose from unchecked climate change. They are inheriting an increasingly damaged planet. Over the six years that I continued to teach atmospheric chemistry and to speak with students about their concerns, my resolve for reducing carbon emissions to avoid dangerous climate change was strengthened. What we are witnessing in the twenty-first century is an intergenerational crime.

PART V

2007–2021

381–410 ppm

Concept 1. "The Atmospheric Thin Film"

Atmospheric Scale height

Integrating

$$\ln p(z) - \ln p(o) = -\frac{M_a \cdot g}{R \cdot T} \cdot z$$

define scale height $H = \dfrac{R \cdot T}{M_a \cdot g}$

$$\Rightarrow p(z) = p(o) \cdot e^{-z/H}$$

$\Longrightarrow$ • Virtually all life on Earth exists below 5 km

• We live in the "Atmospheric Thin Film"

• Our existence depends on its physical & chemical properties

• Altering its composition by adding excess CO_2 and CH_4 changes the properties all life depends on!

CHAPTER 17

THE UN INTERGOVERNMENTAL PANEL ON CLIMATE CHANGE

In the early days at Baring Head, I'd run the atmospheric CO_2 project on a shoestring. Now, things were different. There was widespread support for climate change research by the New Zealand government and, overseas, several international organisations were coordinating global research into climate change issues. These included the highly influential United Nations Intergovernmental Panel on Climate Change (IPCC), set up to provide comprehensive reviews of the state of scientific knowledge of climate change, its social and economic impacts and potential response strategies. And because of the growing importance of changes in air chemistry, an international network of scientists had been created, called the International Global Atmospheric Chemistry project (IGAC).

In 2002 I was invited by IGAC to join their scientific steering committee, which advised on publications and helped organise the programme for their next international conference in 2004. My suggestion that we run the eighth biennial IGAC conference in Christchurch, New Zealand, was immediately accepted. I considered it a huge honour to be serving IGAC beside the world's best atmospheric chemists – the group included people like Paul Crutzen, Uli Platt and Dieter Ehhalt. But how did I get there? I'd never run a conference before and had no idea what I was up against.

I formed a local organising committee of four scientists from TROPAC and the NIWA accountant. Soon we came to be known as 'the tight five', referring to forwards in a rugby scrum. We raised a lot of sponsorship money and hired top experts in conference organising, web development, graphic design and a bewildering range of other fields. A Ngāi Tahu local graciously agreed to open the conference with a pōwhiri.

I'd never tackled a project like this, and in the lead-up I would often lie awake at night wondering whether I'd forgotten some vital detail. But the teamwork of the tight five and the conference organisers carried us through – the event was a huge success, attracting five hundred atmospheric chemists from around the world for a five-day science programme. Many of the scientists and students I'd worked with in Jülich attended, including Uli Platt who had subsequently moved to Heidelberg University. The conference helped cement the close bonds in atmospheric research between Germany and New Zealand. For the conference dinner, the tight five organised entertainment by a Christchurch circus school, with aerial acts, and clowns. A highlight for the scientists was seeing the NIWA CEO, Rick Pridmore, and me carried off by a large person in a gorilla suit – I had no idea this was coming and it scared the hell out of me. The conference marked a coming of age for atmospheric chemistry in New Zealand, showcasing the international status of our work and our significance in cutting-edge climate research.

In 2004 I received a call from the IPCC headquarters in Geneva telling me I'd been selected as a lead author for their fourth assessment report, the AR4. I would be one of fourteen lead and coordinating authors working on a chapter entitled 'Changes in Atmospheric Constituents and Radiative Forcing', for Working Group I of the AR4 report *Climate Change 2007: The Physical Science Basis*.

Irena was blown away when I told her about the phone call, and said what a huge honour the selection was. But I was wary of what the next three years would look like – international travel, an enormous workload and lots of meetings. Our children had left home by this point, and Irena was flat out with work in her own career. She encouraged me to accept the role. Neither of us had any inkling where this was going to lead over the next few years.

The first clue came when I received the formal invitation letter from the IPCC, part of which read:

> Undertaking the author role for which you have been selected is a significant responsibility and will inevitably require a substantial time commitment on your part. However, this is also an opportunity to ensure that the enormous efforts of the climate research community in recent years are communicated as effectively as possible to policymakers. The many significant developments that have occurred since the IPCC's Third Assessment Report and the broader scope planned this time for the WG1-AR4 provide the opportunity to make this the best summary of climate change science ever undertaken. The challenge is large but we think you will agree that the result is extremely important . . .

A significant responsibility indeed, one that I could never have anticipated. Over the next three years, on top of my normal duties at NIWA, the AR4 completely dominated my life and I had to work most weekends and evenings. The IPCC itself does not engage in scientific research; its mandate is to provide comprehensive reports assessing the current status of climate change research using up to date peer-reviewed scientific journal literature. The reports contain in-depth scientific material as well as summaries for policymakers, and have become the most trusted source for worldwide governmental decisions on emissions control targets.

My role in the team was to assess the current scientific literature on increases and trends in atmospheric greenhouse gases like CO_2, methane and nitrous oxide, compare emissions estimates and source information, and provide radiative forcing estimates as well as information on gases which had been identified as important for the Kyoto and Montreal Protocols. I began work for the IPCC immediately after the IGAC conference, travelling to the first AR4 lead authors meeting in Trieste, Italy, and immediately was plunged into the most intensive period of scientific analysis of my career.

The work involved reading and assessing hundreds of recent journal articles, selecting the most important and relevant, and summarising their content in drafts for the IPCC.

For three years I seemed chained to a computer desk day and night. Because the other lead authors were in countries around the world, making telephone calls and writing emails at any time of the day and night became the norm, and I was often working well after midnight. The report was written as a series of four drafts, each of which attracted literally thousands of review comments from various government experts and scientists from all over the world, including one scientist who, bizarrely, turned out to be a climate change denier. I was required to respond to every single comment with notes on what I'd done to address the comment in the report sections I had written. Because of its reputation and huge responsibility, the IPCC receives a lot of criticism and is a constant target for climate change deniers and disinformation campaigns. But I would challenge anyone to come up with a more rigorous, transparent and comprehensive scheme for assessing the best scientific information available on climate change. Any weakness in the reports is rooted out by multiple layers of peer review.

I was working on the IPCC report in mid-2005 when I received a phone call from Peter Guenther at Scripps.

'I have some very bad news,' he said. 'Dave Keeling passed away yesterday – very suddenly in Montana. I thought you'd want to know.'

I was stunned and could only stammer a brief reply. Decades of contact with this wonderful man flashed through my mind. Without his early encouragement and influence, my life would have been completely different. I'd spoken with Keeling only weeks before, talking about his ideas on the airborne fraction of CO_2, a topic I wanted to include in my IPCC report chapter. At this stage we were living at Plimmerton. After the phone call I remember walking among the trees I'd planted, down a path and steps to the beach, where I sat on a rock and looked west across the ocean. I thought of how Keeling had been a presence my entire working life. He had been inspirational, intuitive and dedicated. With his Keeling Curve, he left the atmospheric community an extraordinary legacy, one that will never be taken for granted.

Shortly after this, Keeling's eldest son Ralph – a brilliant atmospheric chemist who had taken over his father's CO_2 programme – asked if I could act as an international scientist to provide independent advice to Scripps on various aspects of their CO_2 research. I counted it as an honour and, although I'd always closely followed the research since embarking on my own path, once again I became closely involved with the Scripps CO_2 programme and its staff, especially Peter Guenther.

As part of this role I attended a conference in Hawai'i celebrating Keeling's work and the fiftieth anniversary of the global CO_2 record in 2007. I was used to such scientific meetings at this point, but this one was very different. As well as climate scientists from American universities and science institutes, the conference organisers including Ralph Keeling had invited a range of non-science delegates including representatives from US coal and nuclear power industries, the military and US federal and state officials. I was one of only two scientists from outside the US at the event.

The climate science presented at the meeting was sobering: massive die-back of coral reefs, diminishing polar sea and glacial ice, and catastrophic loss of ecosystems. This grim forecast was immediately followed by a presentation from Fred Palmer, vice president of Peabody Energy, at that time the world's biggest coal company. He'd just arrived from the mainland in his private jet and lost no time putting up a slide that showed an enormous rampart of coal dwarfing a tiny pick-up truck at its base.

'See this coal,' he said, locking his eyes on the scientists in the front row. 'America needs this and we're gonna burn it!'

Still appalled by slides of dying coral and decimated ecosystems, there was a collective gasp of horror from the audience. I'm still shocked when I think back to those images – I'd liken Fred Palmer's entrance to someone gate-crashing an Alcoholics Anonymous meeting with a couple of bottles of bourbon under their arm. But the conference organisers had done the right thing – understanding begins with contact and dialogue, and this was exactly the space where I felt I should be.

'Dave! Are you awake?' My childhood best friend Con was yelling in my ear. It was October 2007, and I was staying at his place in Christchurch after presenting a climate change talk the night before.

'No, I'm not. Go away!' I grunted.

'Wake up, scumbag! You know that report thing you've been working on for the last couple of years? I just heard on the radio that it's won a Nobel Prize!'

I staggered out of bed and checked my emails. There they were – dozens of messages of congratulations from my co-authors from the IPCC, as well as news reports.

> The Nobel Peace Prize 2007 was awarded jointly to Intergovernmental Panel on Climate Change (IPCC) and Albert Arnold (Al) Gore Jr. 'for their efforts to build up and disseminate greater knowledge about man-made climate change, and to lay the foundations for the measures that are needed to counteract such change.

I had difficulty getting my head round what had just happened. I was a lead author of a report, part of the IPCC process, that had just been awarded a Nobel Peace Prize. Good grief! What did this mean? After a short discussion with my co-authors, we decided that the award should acknowledge the many people, for example librarians and administrators, who had helped us in different ways. The IPCC sent certificates to us and to all the people we nominated. This became one of my proudest possessions.

But on the back of these extraordinary few years, I was facing another personal tipping point in my career. While I had been challenged, rewarded and energised by my work, I had to acknowledge something that – as a lifetime witness of a rapidly changing atmosphere – had become glaringly obvious. The implications of our science, although well publicised, were being ignored. It appeared either that no one understood the consequences of what our data and measurements showed, or that no one cared. Carbon emissions continued to rise, and we continued to do nothing about it.

The activities and profiles of climate change deniers increased. I was heartbroken and frustrated to see the attention their lies received. I wanted to speak out even more, but due to funding constraints NIWA had difficulty supporting public outreach activities like this and I had to do most of it in my own time. After discussion with Irena and other friends, I decided to resign from NIWA and become a self-employed independent scientist working on climate change outreach and sustainability. I was sixty years old, and proud to leave behind a thriving atmospheric chemistry group staffed by brilliant and internationally recognised researchers. Leaving NIWA was a leap into the unknown but the decision felt right.

CHAPTER 18

ALARMISTS VERSUS DENIERS

In the 1970s, when I began my career in atmospheric science, society had had time to plan a gradual phase out of its dependency on fossil fuels. By the early twenty-first century, the situation had become urgent. The science was clear: to avoid even more damage to ecosystems and almost an unfathomable fallout, humans needed to reduce carbon emissions immediately. This was going to involve monumental changes to energy infrastructure, transport and agriculture.

I'd watched the stupidity for long enough. If I was going to increase my public outreach on climate change, I needed a better understanding of energy issues. With this in mind, Irena, Greg and I registered a small private science company, LOWENZ Ltd, aimed at climate change education, renewable energy and sustainability. By this stage Greg had graduated with a degree in computer science and had worked first at Heidelberg University in Germany and then in London before returning to New Zealand. Passionately interested in climate science and renewable energy production, Greg helped us with the installation of small renewable energy systems as well as data analysis for publications and mathematical modelling.

Worldwide, the greatest single source of atmospheric CO_2 is emissions from coal-fired electric generation plants. A societal shift away from this dependency is incredibly difficult but, twenty years ago, Germany was already showing how effective 'distributed generation' renewable energy systems could be. In 1970 about 1% of Germany's electrical energy was renewable. In the new millennium photovoltaic cells were installed on house, farm and small factory roofs all over the country, and farmer cooperatives erected wind turbines. By 2007 Germany's total electricity generation was above 10% renewable. In the first nine months of 2020 it was about 50%. Many of Germany's old low-efficiency brown coal power plants have been phased out, and the country's CO_2 emissions have dropped. They call this the *Energiewende*, or energy transition, of which Germans are justifiably proud. Their example demonstrates what even a highly industrialised country can do to reduce carbon emissions.

After leaving NIWA, I continued to teach atmospheric chemistry at Victoria University of Wellington and to give public climate change and sustainability talks. I also spent several months becoming familiar with small photovoltaic systems. As a trial, I built a small photovoltaic system to power the LOWENZ office and, importantly, our beer fridge in Plimmerton for eight years until we moved to Petone. We also helped others with their own off-grid installations in remote parts of New Zealand. A lot of this was simple education in electrical energy usage. If you live off-grid, the daily energy you generate is a precious resource that needs to be used wisely. It is a wonderful primer on how to live on a finite planet.

Our small company was going well and I enjoyed the challenge of being self-employed. In a small organisation like ours, managing and spreading workload can be a feast or a famine. But Victoria University of Wellington had given me an adjunct professorship in atmospheric chemistry at the Antarctic Research Centre, which kept me in regular contact with their teaching staff and students including Peter Barrett, one of New Zealand's best-known Antarctic scientists.

One day I had a very unusual request from Gareth Morgan, a high-profile economist who had heard about my work with the IPCC and as an independent scientist. Gareth and his wife Jo would rank among New Zealand's more colourful but trusted characters. They are multi-millionaires and run a family trust which grants generous donations to hundreds of charities. Both ride motorcycles and have travelled many parts of the world with their organisation, World by

Bike. From the saddles of their motorcycles they have personally seen the effects of climate change around the world, from the destruction of boreal forest by pine beetles no longer killed in freezing winters to stranded fishing boats 200 kilometres from the now dried-up Aral Sea and widespread desertification in Africa.

Gareth had noticed a lot of shouting back and forth about the reality of climate change, and wanted to get to the bottom of the debate. He wanted to write a book for the layperson about the arguments for either side, and he wanted my help. My first instinct was to tell him there was no debate: climate change was already happening. I wasn't impressed. 'If you live in Alaska, you only need to look out your window,' I said.

Gareth saw that climate change had huge political and economic implications and his aim was to understand more about the science. He insisted that the most effective way to do this was by examining the arguments put up by climate sceptics, many of whom admit that climate is changing but due to 'natural causes'. With co-writer John McCrystal, Gareth decided to examine arguments from scientists who were of the view that humans were damaging the atmosphere against the arguments of those who maintained that nothing untoward was happening. It was, he said, a battle between alarmists and sceptics.

When I told Martin Manning that I was considering the project, his response was immediate. 'You're bloody mad!' he said. 'There is no debate – you're running the risk of being exploited, and it will be a mess if he decides the sceptics are right!' This was a good point. By this point, Martin had become a prominent figure in the United Nations Intergovernmental Panel on Climate Change. Today, quite rightly, he is considered one of New Zealand's top international climate scientists. I discussed the situation with Irena and Greg over the next couple of days. The more we thought about it, the more we realised that this was exactly what we wanted to do with our small business. I had decades of experience in climate science and I'd encountered a lot of climate sceptics along the way. I was prepared to back myself against them.

After making a decision, I met with Gareth and John in Wellington. We had an intense, often terse discussion, but it was clear to me that Gareth was completely genuine. He insisted on getting to grips with climate change science and wanted to present this to the public in an accessible form. But he was adamant that there *was* a debate and wanted the credibility of the sceptics investigated. I became fascinated. I could see that this would become a huge project with high stakes for public outreach and enlisted Peter Barrett and Lionel Carter, a well-known New Zealand oceanographer, to help me. With funding from the Morgan Family Foundation, we embarked on a wild ride of methodically refuting every single climate sceptic's argument that Gareth and John could find.

The most significant of these arguments stemmed from a report by an organisation called the Non-Intergovernmental Panel on Climate Change, headed by well-known climate denier Fred Singer. The fifty-page report, published in 2008 and entitled 'Nature, Not Human Activity, Rules the Climate', intended to point out deficiencies in the AR4, the Nobel Peace Prize-winning report that I had worked on for the IPCC. Carefully going through page after page of misleading arguments, and translating these for Gareth and John, took weeks of painful and frustrating work. Gareth and John were satisfied with our responses and went on to find work from other sceptics to throw at us, a process that lasted more than a year.

A year later their book, *Poles Apart: Beyond the shouting, who's right about climate change?*, was published with a launch at the Parliament Buildings in Wellington. In attendance was a well-known climate sceptic with an engineering background, who in an audience question made a factually incorrect comment on atmospheric radiative heating. I was astonished when this was immediately rebutted by Gareth using validated science from the book. He and John had concluded that 'the science is practically irrefutable', but that a lot more research needed to be done before 'science can claim to understand how the climate really works'. Of course more climate change research is needed, but the urgency of the emergency we face is clear. There is no time to wait.

What did I learn from the process? Above all, humility. As a scientist, it's worth going back to basics and ensuring you are on top of every tiny detail underpinning your area of research. I also learned the importance of being able to communicate science to the public. I've heard some twisted arguments on climate change over the years, but Gareth taught me the importance of listening, evaluating and explaining. His book, and the roadshows that he and Jo ran around New Zealand in 2009, had an impact on the public's perception of climate change in New Zealand. I took a punt working with him and I'm proud to have been part of the process. He still calls me 'the alarmist', which I take as a compliment.

My feelings about alarmism and being an alarmist have changed. For most of my career I have been content to publish my research in peer-reviewed international scientific journals, and write summaries of the results for administrators, science funding providers and policymakers. The feeling I'd shared with colleagues was that if a professional scientist 'meddled in politics', their science would be compromised. But where has that got me and thousands of other climate scientists? A career in climate change science, yes, but almost nowhere in terms of reducing carbon emissions. Decades of research have provided the facts of the climate emergency we face. But although these stark truths are increasingly picked up by responsible investigative journalists, little is being done politically to address the issue at the heart of the climate emergency: that we need to decrease atmospheric carbon emissions, starting now. For many, to be described as an 'alarmist' is derogatory, perhaps implying excessive or exaggerated alarm about a threat. But in the 2020s, when the climate emergency is a proven existential threat to the human race and life on this planet, an alarmist can be an informed person who understands the actual danger ahead and is committed to sounding the alarm.

With a deep rumbling followed by a series of high-pitched cracking sounds, a slab of ice slid from the edge of the glacier ridge into the sea below, causing the eruption of a small wave which jostled the shards of ice below the glacier. The Inuit call it 'white thunder': the sound glaciers make as enormous gravitational forces move them inexorably from higher valleys and mountains to lower regions like the sea. It's a process that has occurred for millions of years as they retreat and advance. Now, human-induced climate change is leading to unprecedented rapid retreat of almost all the world's glaciers.

Irena and I were standing at the bow of the *Safari Endeavour*, a small eco-adventure and research ship owned by a company called UnCruise. A few days before, we had left Juneau, Alaska, where the tiny ship had been dwarfed by multi-decked cruise liners. Our ship was about the size of one of their lifeboats and carried fewer than a hundred passengers and crew. The voyage had no set timetable – we travelled and stopped at spots dictated by what we saw. We had stopped at Glacier Bay National Park and Preserve, Alaska's southeast passage, to watch the almost 100-metre-high face of the Margerie Glacier gradually disintegrate.

One of the most robust projections of climate change science is that polar regions will undergo 'polar amplification', where temperatures in those regions increase at a faster rate than the rest of the planet. This effect is already highly visible in Alaska and Siberia, where record temperatures are documented nearly every year. You don't have to convince Alaskans about climate change – the effects are obvious. We arrived in Alaska's southeast passage at the end of April, a month notorious for cold, overcast days. Yet for the ten days we spent there, we experienced 20°C heat with brilliant sunshine.

In the latter stage of my career, I was appointed by the government's Ministry of Science and Innovation as their New Zealand–Germany Science and Innovation Coordinator. The duties required looking for areas of scientific excellence, promoting them and developing the bilateral science relationship. I worked with ministry officials and science funding agencies in both countries to develop major new bilateral science initiatives, and the passion and willingness of the people I met and worked alongside restored my faith in what can be achieved at a government level. One of my last visits as coordinator was to the same atmospheric chemistry institute in Jülich where I had completed my PhD in the early 1980s. It was autumn and the trees in the forests around Jülich were dressed in glorious autumn colours. I marvelled at the beauty of it, the crisp air and brilliant blue of the sky above the lakes, the tree-covered hills of the Eifel. I was wearing a suit and entering the research institute in a chauffeured car as an official representative of the New Zealand government. The last time I'd been there I'd left wearing scruffy jeans and riding my old bike.

I was to make one other important homecoming. It was 2015, and I looked out across an auditorium at Scripps, California, at friends and colleagues I'd known my whole working life. Their warmth and encouragement was overwhelming and I was in tears. Ralph Keeling had asked me to give a presentation in honour of his father and my mentor, Charles David Keeling, at a symposium celebrating the recognition of Scripps by the American Chemical Society as a historic site for its discovery of what is now known as the Keeling Curve. I'd begun working at Makara at the age of twenty-three. Forty-five years later, I was able to put into perspective my journey with the atmosphere – the times of elation, the dark places too, and the privilege of knowing and being mentored by legendary scientists like Dave Keeling, Athol Rafter and Dieter Ehhalt.

When I began my research, I was alone. These days, no climate scientist is. Worldwide, there are tens of thousands of scientists working on different aspects of climate change. The validity or worth of their research is rarely questioned. My visit to the Scripps symposium was one of pride and honour. Our research was summarised perfectly by Ralph Keeling, who wrote, 'For the Mauna Loa and other long-term atmospheric CO_2 records, the point of diminishing scientific returns has never been realised.'

On our trip to Alaska, Irena and I had paddled kayaks around islands and small icebergs. Bald eagles flew overhead and we saw bears, otters and whales. In a luminous arc the atmosphere merged into the sea, with iridescent colours on the horizon. At night, the stars lit up the faces of glaciers rising above the dark sea below. Beaches and inlets with dense evergreen forests reached down to the water's edge – everything was pristine and whole. Evidence of human life was missing: there were no jet trails overhead, no buildings or roads. It was beautiful. Like parts of New Zealand, much of Alaska is unspoilt wilderness.

We can consider Baring Head and other long-established atmospheric stations as priceless jewels – they have provided irrefutable evidence of the damage humans have caused the atmosphere over decades. They also tell us exactly what we need to do in order to take heed of the warning and make change. The question is: will we?

ONE ATMOSPHERE, ONE DECADE, ONE LAST CHANCE

The small streets, houses and cafés in Petone, this quirky little suburb of Wellington, New Zealand comfort me. It's home. The local dairy owner, the Dandelion Café staff round the corner, the greengrocer, the pharmacist, the bookseller and organic food store owner all know me. I love the daily greetings, the feeling of being known and knowing. I love the warm and dry cottage I share with my wife, Irena, its back-garden vegetable plot and small greenhouse and the friendship with our neighbours over the back fence. And I love the daily routine of getting on my bike and going for a ride along the Wellington Harbour foreshore and the local river bike trails. A couple of times a week, I hop on a bus or a train for the twenty-minute ride into Wellington for meetings or to touch base with our son or daughter. I'm getting older now. Gone are my wild surfing days and weeks spent hiking back country trails.

Petone is just a few kilometres from Gracefield, the Wellington suburb where fifty years ago I began work with Keeling and Scripps on the first continuous atmospheric carbon dioxide measurements in the mid latitudes of the southern hemisphere. In 2021 atmospheric CO_2 at Baring Head will approach 411 ppm, 90 ppm higher than when I first began measuring it and more than 100 ppm higher than it was in the year I was born, 1946. This increase of more than 30% marks a huge negative change to the properties of the atmosphere on which all life on Earth depends. How do I feel about this? Disappointed, certainly, but also frustrated and angry.

What more could I have done? Like many people, Irena and I have always been aware of the environment we live in and have considered our personal carbon emissions. We moved from Plimmerton into the Petone cottage about five years ago, choosing a location close to public transport, and double-glazed and extensively insulated the cottage to the point where it requires little heating in winter. We have a small vegetable garden and greenhouse, and we grow a lot of our own food. If we have to drive, we use an electric car; otherwise we ride bikes or walk to local shops and businesses. This has required some sacrifices and financial costs. Downsizing to a smaller house can be a painful process – but we enjoy our lifestyle and our carbon footprint is small. We have reduced air travel, a significant source of carbon emissions globally. If we have to fly, we offset our emissions using an audited tree-planting scheme, where long-lived native trees are planted and protected in perpetuity by legal covenant. We're not alone in our efforts. Millions of people worldwide have changed their lifestyles to dramatically reduce their personal carbon emissions.

Individual action is a huge help, but is it enough? Unfortunately, no. Public attitudes and willingness to change lifelong habits help to drive improvements, but the real breakthroughs have to come from the concerted coordinated actions of governments worldwide. By 2017, the Paris Agreement, to combat climate change by accelerating and intensifying the actions and investments needed for a sustainable low-carbon future, had been ratified by most countries. Its goal is to hold global temperature rise well below 2°C this century and to limit temperature increase even further to 1.5°C. And yet almost nothing has changed. The emissions pledges are non-binding and have, in most cases, not been adhered to. Under the Trump administration, the US announced withdrawal from the Paris Agreement, a decision that has been reversed under the Biden administration. If the plans of the Biden administration to invest US$2 trillion in post-Covid green stimulus packages are realised, the US will overtake the EU as the biggest investor in a low-carbon future. This would be transformative, and a very important signal to other countries. Economists have shown the large-scale positive effects on employment and job quality created by a dynamic green economy. Perhaps the US could begin a race to the top, leading the way with a 'green recovery'?

Some countries have made progress. Uptake of renewable energy in Germany has dramatically reduced the country's CO_2 emissions from brown coal power plants. Scandinavian countries have been very effective in reducing carbon emissions, with cities like Copenhagen already close to net zero carbon. But these are exceptions, and not nearly enough is being done to slow current worldwide emissions to achieve net zero carbon by 2050. Where does this leave me? Am I a pessimist or an optimist?

When it comes to the political will and leadership needed to drive the world towards a sustainable future, I'm a pessimist. Time and time again, I've heard rhetoric from politicians focusing on short-term goals at the expense of planning for the future. In 2021, the mainstream media promote responsible journalism and take a hard line with climate deniers. Many journalists hold governments to account over climate change goals. However, hard scientific data is often still manipulated and cherry-picked by politicians. I've spoken to many and liken the experience to walking through treacle. Does their bland decision-making have to do with the structure of democracy itself, with its short electoral terms and lack of incentives for incumbent politicians to make hard and binding decisions for the decades ahead?

As I look around and see New Zealand's highways, jammed with huge diesel trucks and ever-increasing numbers of petrol-powered SUVs and cars, I feel dread. It doesn't have to be this way. What is it about living on a finite planet that humans either don't or won't understand, after all the studies and warnings show that continuing in this way leads to the inevitable collapse of the planet's ecosystems?

When you look at the true cost of the damage to the atmosphere, politicians' claims that action on carbon reduction is too expensive become bizarre. When we burn fossil fuels, we've never factored in the ultimate cost of the damage to the atmosphere caused by excess CO_2. In many countries, if you pollute a waterway, you have to clean it up or pay a substantial fee for the damage – that cost has to be factored in to the cost of running your business. In the case of emitting CO_2 into the atmosphere, you can do that for little or no up-front and immediate cost. Are we offended by people polluting waterways because it is literally in your face whereas CO_2 is a transparent gas? We now know, more than ever, that it hasn't disappeared. The excess CO_2 emitted in the Industrial Revolution is still in the atmosphere all around us. Every year since then, human actions have added an increasing burden of excess CO_2 to the atmosphere, a gas that will remain there for thousands of years and drive an enhanced greenhouse effect. If we continue business as usual, the warnings from the IPCC are clear: we will drive the planet into dangerous climate change, probably within our own lifetime and certainly within the lifetime of our children.

In 2018, economists William Nordhaus and Paul Romer were awarded the Nobel Memorial Prize in Economics for their work on economic models such as carbon taxes to study global problems like climate change. To me, paying a tax or a fee that represents the damage you do to the atmosphere through carbon emissions is such an obvious tool for real change. But many politicians fear they will lose an election over the word 'tax', so are we, their constituents, to blame for demanding lower taxes?

New Zealand adopted an emissions trading scheme in 2008, which was a major step in the right direction to tackle increasing carbon emissions. But the scheme has been so watered down that it seems ineffective; the country's emissions continue to increase, with the fastest growth in the transport and agricultural sectors. On a more positive note, at the end of 2019 the country adopted a Zero Carbon Bill that set climate change targets into law. Shortly after this, to counter growing transport emissions, the government proposed a 'feebate' scheme to lower the cost of low emission vehicles like electric cars. My elation was shortlived: in mid-2020 the idea was ruled out in a political disagreement and disappeared off the radar. And so, my pessimism about political action continues.

But in a swing against the rising tide of populism worldwide, towards the end of 2020 Jacinda Ardern's Labour government was re-elected with an increased majority. In December 2020 her government declared a climate emergency, pledging immediate action to tackle the crisis. The announcement has been backed by a suite of positive measures aimed at reducing emissions, including decarbonising the public sector by 2025. This will involve switching the government fleet to electric vehicles, replacing coal boilers with cleaner alternatives, transitioning to energy-efficient buildings and offsetting remaining emissions in a way that creates a direct financial incentive for the government to reduce emissions. I'm elated by the declaration – it's the first time in my decades-long journey of watching the atmosphere deteriorate that I have seen such a positive move by a government in New Zealand. It's an excellent initiative and in line with countries like Canada, France and the UK, who have taken the same course to focus efforts and tackle climate change.

For most of the last few decades I have been disappointed with the lack of action on carbon emissions reductions by politicians. But on the other hand, I'm very optimistic when it comes to the extraordinary ingenuity of human beings. We already have the tools to combat climate change. The last two decades have seen massive advances in renewable energy electricity generation to the point where these sources are now cheaper than equivalent coal-burning power plants, even before the cost of damage to the atmosphere is taken into account. The International Energy Agency (IEA) reported that, in 2019, almost 30% of OECD electricity was met by renewable sources including hydro, solar, wind, biomass and geothermal.[19]

Crucial to the urgent transition towards a low carbon future will be the skills and experience of engineers. Over the years I've spoken to many groups of engineers, including oil and gas engineers, about climate change. You'd think that a climate scientist talking to a gas engineer would lead to an argument, but that has not been my experience. Part of this has to do with my unusual background – a New Zealand climate scientist with a knowledge of electrical engineering and a German PhD in atmospheric chemistry. I also have an appreciation of the incredible skills held by many engineers in the fossil fuel industry. After all, they are using their talents at our call – they are providing products that the world cannot get enough of. But those same gas and other engineers who have been so maligned by the green movement have the vital skills needed in a new sustainable economy. Their skills are transferable to an economy making widescale use of 'green hydrogen', for example. Green hydrogen, produced by electrolysis of water using excess electricity derived from wind and other renewable energy sources, is already being used in steelmaking, energy storage and transport in Germany and a number of other countries. In New Zealand, the first heavy haulage hydrogen-powered trucks are due to be released in 2021, and a New Plymouth company, Hiringa Energy, plans to have a fleet of 1500 on New Zealand roads by 2026. The hydrogen is used to run electric fuel cells, providing the trucks with a driving range of 500 kilometres and a refuelling time of 20 to 30 minutes, similar to current diesel-powered trucks. With government support, the hydrogen refuelling sites will be installed by local energy company Waitomo.

When I talk to people about this technology and its possibilities, they are astonished. They wonder why they have never heard of it. It's worth checking the IEA and other trusted websites, like CleanTechnica, for updates on the rapid and widespread developments in these fields. Hydrogen fuel cell technology has been around a long time – I remember first seeing it decades ago. Why hasn't it been used? Several reasons come to mind, including conspiracy theories about the oil companies, but to me there is a simple answer. It's because products made from fossil fuels appear to be so much cheaper than sustainable alternatives; the true cost of the climate emergency is never factored in when the products are sold to customers.

So what is the true cost of the damage to the atmosphere when you emit a couple of tonnes of CO_2 into it, perhaps during a longhaul flight between Auckland and London or by running a diesel-powered SUV for a year? There are a lot of different answers to that question depending on whether you ask an economist, politician, engineer or a climate scientist. If you ask a chemist how, and how much it would cost, to remove a tonne of CO_2 from the atmosphere, they would probably throw up their hands in horror, come up with a figure of $1000 per tonne and a very complex apparatus. A climate scientist would reply to the question with another, like, 'How much do you think the 2020 wildfires in Australia, California, Colorado, Siberia and the Arctic cost?' And a New Zealand economist would quote the current carbon price on the New Zealand emissions trading scheme site, which in early 2021 was about $37 per tonne. To me that sounds ridiculously cheap, measuring in crude economic terms the cost of the damage by carbon emissions into our only atmosphere.

We've been blinkered into thinking that there are no alternatives to fossil fuels for running an economy and society. But engineers and economists can point to several alternatives, and we need to adopt the ones that provide a sustainable future in this decade. A new field has emerged which has come to be known as 'transition engineering', where engineering and scientific principles are used to provide systems which do not compromise the ecological, societal and economic systems that future generations will depend on.

Engineering solutions will be especially valuable in tackling the rapidly growing emissions from transport. Worldwide, liquid fuels like petrol and diesel for cars and trucks, jet fuel for aviation and bunker fuels for shipping accounted for more than 20% of total CO_2 emissions in 2016.[20] Growing at a faster rate than any other sector, transport poses a major challenge to reducing emissions in line with the Paris Agreement. To keep global temperature rise within a range that averts the worst climate impacts, IPCC and other climate modelling show transport emissions must decline. Transitioning to zero-emission transport is crucial. Solutions include clean fuels, improved vehicle efficiency, changes to how we move people and goods, and building sustainable cities.

Electrification eliminates tailpipe emissions of CO_2 and particles that damage our lungs. It harnesses the potential to decarbonise the power grid. In New Zealand there has been an encouraging uptake of electric vehicles: the number has trebled over the last two years, but from a very small base. In 2020 about 0.5% of the country's vehicle fleet was electric.[21] In Norway, by mid-2020, about 14% of the entire vehicle fleet was electric, and in the last two years between 50–60% of all new cars sold have been electric.[22] Norway's government has incentivised the use of electric vehicles and plans to stop sales of diesel and petrol cars by 2025. Because Norway's electricity grid is almost entirely powered by hydro schemes, the drop in transport carbon emissions has been dramatic with the added benefit of reductions in city air pollution.

New Zealand's population is similar to Norway's, and its electricity is about 85% renewable with plans to increase this to 100%. If we follow Norway's example and transition the vehicle fleet to electric, we can reduce the country's CO_2 emissions, improve air quality and decrease dependence on imported oil all at once. In a fossil-fuel powered car, barely 25% of the energy in the fuel is transferred into motion, a fraction that can never be significantly improved due to basic physics and the second law of thermodynamics.

But the same fundamental restrictions do not exist for an electric car: more than 90% of the energy in its battery is converted into motion. Simple engineering calculations show that, without modification, the existing New Zealand electricity grid could easily power a fleet of two million electric cars travelling an average of 12,000 kilometres per year. The extra electricity required would be less than 10% of current total production, and no new power stations would need to be built. The savings in CO_2 emissions would be a massive – five million tonnes every year. New electric vehicles arriving in New Zealand are relatively expensive, but cheaper secondhand ones are widely available. Incentives to make electric vehicles cheaper are straightforward, what economists refer to as 'low-hanging fruit'. So far, the only thing preventing New Zealand from following Norway's example on cars is politics.

I've mentioned two major sources of CO_2 emissions – electricity generation and transport – and how these emissions can be or are being reduced. But there are many other culprits, as almost every aspect of modern life involves burning and using fossil fuels. In the last seventy years we have increased worldwide energy consumption from the equivalent of 20 TWh in 1950 to about 140 TWh in 2017, a factor of sevenfold in what has become an extraordinary and dangerous addiction.

There is no doubt that reducing carbon emissions to avert disastrous impacts of climate change will be a gigantic undertaking. No single solution to this problem exists. It will require concerted effort from all parts of society, above all governments, but also engineers, scientists, economists, teachers and farmers. We can feel optimistic of the rapidly emerging technologies available to help reduce carbon emissions, among them hydrogen generation and storage from surplus electricity, synthesis of sugars from CO_2 and water, information and nanotechnology, bioengineering and educational science to name a few. The challenges ahead are formidable but I truly believe that, given the will and with concerted action, human beings are more than capable of building a sustainable future.

Above all, I am optimistic when I see the effectiveness of young people like Greta Thunberg, the Swedish schoolgirl who started the Fridays for Future movement, uniting millions of mostly young people whose climate strikes raise awareness of the emergency from which their generation has the most to lose. Their target audience is politicians, and they are having an impact – after all, their parents are voters and soon they will be too. In early 2021 I was proud to address a large group of young strikers outside New Zealand's Parliament Buildings. The people I spoke to were highly motivated, passionate and focused on change. They understand that coordinated, effective multinational government action using all the embedded skills of the human race is what is needed now to change a future that, if we don't act quickly and decisively, will become inevitable.

We only have a decade to change our behaviour. I've watched the atmosphere for most of my life. It's the only one we have.

COVID-19, URGENCY AND POLITICAL WILLPOWER

Within the horror of the Covid-19 pandemic has emerged one small silver lining with a very obvious lesson. For the first time in decades, there has been a significant drop in carbon emissions: for 2020 these were estimated to be about 7% lower than in 2019, dropping to levels last seen in 2010.[23] With international aviation all but shut down and a dramatic drop in road traffic and worldwide shipping, the global price of oil has plummeted due to lack of demand.

For a brief moment we witnessed the benefits of a world without the pollution associated with transport. With vehicles, noise and exhaust fumes absent, city streets became pleasant walkways. During New Zealand's strict lockdown between April and May 2020, cleaner air was very noticeable, and I was one of many who enjoyed biking and walking on city streets with little or no traffic. In Wellington, native birds began to recolonise empty central city streets – a rare native falcon even terrorised local flocks of starlings. Many people worked from home, with their employers reporting an increase in productivity. After the nationwide lockdown ended and the New Zealand domestic economy reopened, there are definite signs that work habits have changed – many people will now work from home more regularly and there has been a drop in peak-hour traffic. People are living locally, buying locally and supporting neighbourhood businesses to the extent that some small businesses are reporting increased sales. Many New Zealanders are questioning a return to life pre-Covid-19 – this crisis has forced a rethink of values and personal choices. Worldwide, there has been an increase in trust for coordinated global action and science through groups like the World Health Organisation and the United Nations.

The Covid-19 crisis has demonstrated that governments can respond decisively to emergencies when public support is present. This has been especially true in countries with strong trusted leadership acting swiftly under advice from scientific and medical authorities. Resolute intervention has stabilised infection rates, prevented health systems from being overwhelmed, and saved lives. Countries with efficient centralised medical care and trusted experts have done well; countries with populist governments and a distrust of experts have tended to do poorly. Ironically, countries that believed they were favouring their economies over health by refusing to implement measures like strict lockdowns have fared worse economically in the long-term as Covid-19 case numbers and deaths continued to soar. Early in the pandemic, the New Zealand government was criticised by many for enforcing one of the strictest lockdowns in the world, essentially shutting down the economy for all but essential services like food supply. But by the end of 2020, New Zealand's domestic economy was fully open and with the impacts of those strict measures far less severe than projected by bank economists. The New Zealand government's measures, which were seen as harsh at the time, have instead allowed the economy to prosper. The country is faring better than most as we enter the recovery phase after the pandemic, with several promising Covid-19 vaccines becoming available from overseas.

Although the timescales are very different, economists have shown that allowing carbon emissions growth to continue will have a devastating effect on future economies when the more dangerous events of climate change become commonplace. Already, extreme climate events in the form of hurricanes, wildfires and heatwaves have huge effects on economies around the world. It's clear that taking action now in reducing carbon emissions will reduce the economic effects of climate change too.

Humans are facing multiple existential threats: the climate

emergency, pandemics, rising inequality and the failure of democracies, to name a few. The links between these threats are clear. With the deteriorating health of the planet and the loss of natural ecosystems, our increasing human population densities have magnified the chances of viral transmissions from animals to humans, with terrifying consequences like the Covid-19 pandemic. In the case of both Covid-19 and climate change, health – including the health of our planet – and the economy are inextricably linked. A root cause of these threats lies in the behaviour of humanity itself.

The climate emergency is much like the Covid-19 pandemic, except that it's slower and much more serious. Both climate change and Covid-19 involve complex science, resilience, political leadership and action that depends on public support. As I've pointed out, government intervention is necessary to reduce emissions by moving transport, industry and energy generation towards lower carbon and competitive methods of production.

New Zealand's success in tackling the pandemic has been due to a number of factors, including expert advice being acted on decisively. To me, one essential factor is what the prime minister called 'the team of five million'. It was the country's population buying in to what were at times strict rules to overcome the worst effects of the pandemic. The results speak for themselves: the economy is bruised but recovering well, and the death rate is five per million population, compared with well over a thousand per million population in early 2021 for the USA and many European and South American countries.[24] Yes, New Zealand has a huge advantage with its geography, but that does not alone explain the factor of more than 250 in death rate – the difference is almost certainly due to a variety of factors, among them gross mismanagement and public disdain towards 'rules'. Does this come down to differences in beliefs about fairness and appreciation of the rights of the individual versus the needs of the many? It seems to me that for the earth to support a human population projected to rise to 10 billion by 2050, human passengers on Planet Earth will have to abide by 'rules' of sustainability, including living a net zero carbon lifestyle.

How governments use fiscal Covid-19 recovery packages after lockdowns end will be crucial. Public support for action on climate change was increasing before the outbreak of the pandemic; industry and government action was also gathering momentum. Economists are currently advising on stimulus policies that will deliver large economic gains quickly. These recovery packages could help drive the global economy towards net zero carbon emissions or, wrongly used with poor leadership, they could lock us into a carbon emissions nightmare from which it will be nearly impossible to escape. Our twentieth-century economy was largely built on cheap oil, coal and gas whose environmental impacts were never costed. After Covid-19, using public money to support ailing fossil fuel industries would be a disaster.

The terrible months of 2020 have shown that, when facing an obvious existential threat, humans can rapidly reduce carbon emissions. When we do overcome Covid-19, we must not return to a world of ever-increasing carbon emissions. Adversity should lead to strength, resilience, change and understanding. Will we leap from a Covid-19 frying pan into a climate change fire, and squander the opportunities created by emissions reductions in 2020? Or will we use the lessons learned to achieve net zero carbon in the years ahead? New Zealand called on a team of five million to beat the Covid-19 pandemic. In the 2020s, humanity must call on a global team of eight billion to tackle carbon emissions, starting now. There is only one right way forward.

AUTHOR'S NOTE

The current dictionary definition of 'alarmist' is 'someone who exaggerates a danger and so causes needless worry and panic'. Climate change deniers often use this word to put down climate scientists whose task is to understand the forces that drive the global climate, the intention being to suggest that the issues raised are of no real importance. However, when climate scientists like me make statements that sound 'alarmist', they do so because they are genuinely alarmed by the implications of their measurements of a rapidly deteriorating planet.

I've struggled with the title of this book since it was first suggested to me by my publisher, Fergus Barrowman. I hated the idea at first, but realised that 'alarmist' in this context introduces a sense of irony that turns deniers' views against them. We are facing a climate emergency, and more and more climate scientists are sounding the alarm. We are alarmed by the damage we are recording. There is little time left, and it is essential that the global population shares that alarm and supports action to stop further harm being done.

*

In 2019 I was contacted by a New Zealand journalist, Joel MacManus, who asked whether he could interview me for an article about my life as an atmospheric chemist. He'd heard that I was the scientist who began the Baring Head programme. I agreed and he went on to write an article entitled 'Dave Lowe found measurable proof of climate change 50 years ago: He's watched in horror ever since'. His full-page article, with pictures of Baring Head, various labs where I've worked, and me as a teenage surfer in Taranaki, were printed by *Stuff* in many newspapers around the country and posted on their website. Joel told me that the article had more hits on *Stuff*'s website than any other article that month, and he subsequently went on to win a media prize for the piece. You can find his and other articles about my thoughts on the climate emergency by googling 'Dave Lowe NZ scientist'.

Immediately after Joel's article was printed, I began to receive a flood of text messages, phone calls and emails from all over New Zealand and Australia asking me for information and to give presentations about my life and my thoughts on climate change. His article had engaged readers and made them act. Why did it have such a wide impact? I think there are several answers, not the least of which is the brilliance of Joel's writing. But it was also the subject matter – climate change has become a topic that many people are concerned about. And his readers could identify with a story about an ordinary New Zealander who has done his best for fifty years on a journey with many twists and turns.

Before Joel contacted me, I'd already thought about writing some kind of memoir about atmospheric CO_2 and Baring Head. But I had no idea how. I'm just a scientist, used to writing scientific journal articles, not books. Following advice from an informal Wellington writers' group, I began writing the story in the form of my journey, the one you've followed through these pages. During the writing I've been fortunate to have access to literally hundreds of old letters, diaries, reports, funding proposals and photographs, all of which have helped me piece together important parts of the story, laying down facts as well as a timeline.

As part of this work, I was contacted by Costa Botes, a talented filmmaker who worked with Peter Jackson on *The Lord of the Rings* behind-the-scenes documentaries. During 2019 and 2020 he filmed interviews and locations important to my journey, including Taranaki, Baring Head, Makara and Scripps.

Thank you for following me on my journey. Remember that all of your actions count; your buying and travel decisions, who you vote for, and your lifestyle choices – to plant a garden is to believe in tomorrow. We can do this.

ACKNOWLEDGEMENTS

A lot of people helped with the preparation of this book. Early encouragement came from New Zealand author Lloyd Jones, who told me to write about 'tipping points' that changed my life, including surfing. So, I did. I started the process by taking an excellent creative writing course at Victoria University run by Susan Pearce, and soon discovered that I was the only person in the class who had never written stories before! This kicked off a very steep learning curve, where the long-suffering Wellington writers' group I belong to beat my miserable creations into shape. Thanks, in particular, to Fiona Lincoln for introducing me to Joseph Campbell's 'hero's journey', which at first I discounted as a mythical structure of no use to me. But Fiona was right – my fifty-year journey with the atmosphere, through many dark places as well as times of elation, fits the format. And thanks to Vivienne Merrill and Jeni Bryant for advice on early drafts. Thanks also to Canadian author Catherine Cooper and her timeline suggestion which helped me frame the story.

And to Maggie Dyer and librarians everywhere: you are stars! Without Maggie's help I would never have found the precious archived records of the early days of the Baring Head project. The story in these pages is based on those records and other archived material I found while researching the book.

I'm especially grateful to the Keeling family – Louise, Ralph, Drew and Eric – for their encouragement, help and inspiration. Dave Keeling was an early mentor and inspired me as a junior scientist in ways that I am still discovering. It's been a privilege to write about my interactions with him and his early CO_2 measurements, which underpin so much of our knowledge about climate change.

Germany – what a super country! Fabulous people and a fascinating culture, with a superb education system that turned a naïve young Kiwi scientist into someone who knew a little about atmospheric chemistry. Special thanks to my PhD supervisor Dieter Ehhalt, who helped make that happen. From day one in Germany I was helped by Ulrich Schmidt; he has a special place in my journey. And many thanks to Uli Platt for his constant encouragement, sense of humour and science advice. And to our friends throughout Germany, thanks for your friendship and help during the years we spent in your wonderful country. I am very grateful to the Wellington Goethe Institut for their offer of partial support for a writer's residency in Germany to help with the book preparation. Of course, along came Covid-19 and I could not take it up. Thanks especially to Judith Geare and Christian Kahnt at the New Zealand Goethe Institut for encouragement and advice.

I currently play for a blues/jazz band called Mana Blues. My gratitude to the musicians in the band, Colin Bleadsdale, Paul Goodhead and John McKoy, for their continuing support, advice, good humour and encouragement throughout the book project.

For tireless work editing my first rough drafts of the manuscript ready for the team at Victoria University Press, thanks to Con Jackson. Grateful thanks to the VUP team and in particular to Holly Hunter, whose brilliant editing skills have polished a rough manuscript from a scientist. And many thanks to Fergus Barrowman for his encouragement and for taking this project on, thanks to Ashleigh Young for her advice, and to Rebecca Priestley for introducing me to VUP.

Hey Peter, my dear Californian friend – look at what we started all those years ago! Thanks for your friendship and advice over the decades since our first experience of the southerly winds roaring up the cliff at Baring Head.

The fifty-year record of CO_2 is a priceless heirloom from giants like Dave Keeling and Athol Rafter. That legacy continues at Baring Head, with the station being well supported by NIWA and expertly run by their TROPAC group. Very special thanks go to Gordon Brailsford, whose dedication and skills keep the station and its international connections running.

GLOSSARY

AMS
Accelerator mass spectrometry: a technique used for radiocarbon-dating tiny carbon samples, modified for use in atmospheric chemistry.

carbon-13
A naturally occurring stable isotope of carbon used to 'fingerprint' sources of atmospheric gases like CO_2 and methane.

carbon-14
A naturally occurring radioactive isotope of carbon used to track carbon emissions in atmospheric gases like CO_2 and methane.

calibration gases
Also known as CO_2 reference gases, these gases were produced in Dave Keeling's laboratory at Scripps, stored in high-pressure gas cylinders, and shipped to New Zealand. At both Makara and Baring Head, CO_2 in ambient air at the site was compared with the CO_2 values assigned to the calibration gases. Both the CO_2 in the calibration gases and the air sampled at the sites was measured continuously in the same infrared analyser, providing the first measurements of atmospheric CO_2 in New Zealand. The atmospheric data I produced for Makara and Baring Head were critically dependent on the values supplied for the calibration gases from Dave Keeling's lab, and were initially locked to a scale known as '1959 Mano'. Over time, more refined techniques from cooperating international laboratories have improved the atmospheric CO_2 calibration scale, and it has been adjusted upwards slightly from Keeling's original estimates. I have used the original values I measured fifty years ago in the text, but the ppm numbers used in the page headers follow the latest agreed international values provided by NIWA in early 2021. This leads to a small discrepancy but, sadly, the awful discovery I made fifty years ago that CO_2 was increasing around New Zealand remains.

Clean Technica	A trusted US-based website specialising in the latest developments in renewable energy and electric car production.
CRIs	Crown Research Institutes in New Zealand are corporatised crown entities responsible for scientific research that benefits the country.
DSIR	Department of Scientific and Industrial Research
HPLC	High Performance Liquid Chromatography is a widely adopted analytical technique used to identify chemical constituents in liquids, such as traces of performance-enhancing drugs in urine samples from athletes.
ICH3	Institute for Atmospheric Chemistry, KFA, Jülich
IEA	International Energy Agency: an autonomous organisation providing data, policy recommendations and real-world solutions to help countries secure sustainable energy supplies. It's a conservative trusted source for trends in all forms of energy generation, including renewable.
IPCC	The Intergovernmental Panel on Climate Change is an intergovernmental body of the United Nations charged with providing objective scientific information on the risks of human-induced climate change, with relevant political and economic impacts and possible options for response.
INS	Institute of Nuclear Sciences, a division of the original DSIR
KFA	Kernforschungsanlage, Jülich
Kyoto Protocol	An international treaty committing governments to reducing greenhouse gas emissions. It was adopted by about 190 countries in 1997 and came into force in 2005. Countries agreed to reduce emissions over a series of commitment periods by reporting validated progress on emissions reductions.
Mass specs	Analytical laboratory equipment used to make precise measurements of stable isotopic ratios in elements like carbon and oxygen.

Montreal Protocol	An international treaty requiring the phasing out of substances that cause destruction of ozone in the stratosphere. First ratified in 1987 by 196 countries, the agreement is credited with halting the increase of dangerous holes in the ozone layer appearing over Antarctica every southern spring.
Net zero carbon	A situation where human atmospheric carbon emissions balance carbon removal processes and the goal of the Paris Climate Agreement for sustainability.
NCAR	National Center for Atmospheric Research
NIWA	National Institute of Atmospheric and Water Research, one of New Zealand's CRIs
NOAA	US National Oceanic and Atmospheric Administration
Paris Agreement	Also known as the Paris Accord, this agreement was negotiated by 196 countries and adopted by consensus in 2015. Its goal is to keep the increase in global average temperature well below 2° above pre-industrial levels but to limit the increase to 1.5°, recognising that this will substantially reduce the risks and impacts of climate change.
ppm	parts per million
ppt	parts per trillion
Scripps	Scripps Institution of Oceanography, La Jolla, California
TROPAC	Tropospheric Physics and Chemistry is the group at NIWA that I worked for and which now maintains the priceless Baring Head CO_2 and methane records.
WMO	World Meteorological Organization, based in Geneva. It and the United Nations Environmental Programme spawned the IPCC, which was the brainchild of Bert Bolin, a Swedish meteorologist and early collaborator of Keeling's in the 1960s.

ENDNOTES

1 Oliver Milman, 'California fires set bleak record as 4m acres destroyed', *The Guardian*, 5 October 2020, theguardian.com/us-news/2020/oct/05/california-fires-4m-acres-wildfires-burn

2 Sarah Gibbens, 'Plastic bag found at the bottom of World's Deepest Ocean Trench', *National Geographic*, 3 July 2019, nationalgeographic.org/article/plastic-bag-found-bottom-worlds-deepest-ocean-trench/

3 Melanie Bergman et al., 'White and wonderful? Microplastics prevail in snow from the Alps to the Arctic', *Science Advances* 5:8, 14 August 2019.

4 Svante Arrhenius, 'On the Influence of Carbonic Acid in the Air upon the Temperature of the Ground', *Philosophical Magazine and Journal of Science* 5:41, 1896, pp. 237–76.

5 BP Statistical Review of World Energy is a series of annual reports freely available online at bp.com/en/global/corporate/energy-economics/statistical-review-of-world-energy.html. The reports record usage of fossil fuel, renewable and nuclear energies. Note that there are many other sources of this data, for example, ourworldindata.org.

6 'Restoring the Quality of our Environment', *Report of the Environmental Pollution Panel President's Science Advisory Committee*, The White House, 1965.

7 Charles David Keeling, 'The Concentration and Isotopic Abundances of Carbon Dioxide in the Atmosphere', *Tellus* 12:2, 1960, pp. 200–03. Further, you can explore an excellent article on the significance of the Keeling Curve by Euan Nisbet, 'Cinderella Science', *Nature* 450, 2007, pp. 789–90.

8 D.C. Lowe, 'Atmospheric Carbon Dioxide in the Southern Hemisphere', *Journal of the Clean Air Society of Australia and New Zealand* 8:1, February 1974, pp. 12–15.

9 D.C. Lowe, P.R. Guenther and C.D. Keeling, 'The Concentration of Atmospheric Carbon Dioxide at Baring Head, New Zealand', *Tellus* 31, 1979, pp. 58–67.

10 D.C. Lowe, U. Schmidt and D.H. Ehhalt, 'A New Technique for Measuring Atmospheric Formaldehyde', *Geophysical Research Letters* 7:10, 1980, pp. 825–28.

11 D.C. Lowe and U. Schmidt, 'Formaldehyde (HCHO) Measurements in the Nonurban Atmosphere', *Journal of Geophysical Research*, 88:C15, 1983, pp. 10844–858.

12 D.C. Lowe, 'Atmospheric Effects of Nuclear War', *New Zealand Journal of Science* 27, 1984, pp. 317–26.

13 D.C. Lowe, C.A.M. Brenninkmeijer, M.R. Manning, R. Sparks and G. Wallace, 'Radiocarbon Determination of Atmospheric Methane at Baring Head, New Zealand', *Nature* 332, 1988, pp. 522–25.

14 Christopher Cantrell et al., 'Carbon Kinetic Isotope Effect in the Oxidation of Methane by the Hydroxyl Radical', *Journal of Geophysical Research* 95: D13, 1990, pp. 22455–462.

15 C.A.M. Brenninkmeijer et al., 'Interhemispheric asymmetry in OH abundance inferred from measurements of atmospheric ^{14}CO', *Nature* 356, 1992, pp. 50–52.

16 Sharon Begley, 'The truth about denial', *Newsweek* 150:7, 2007, pp. 20–7, 29.

17 Naomi Oreskes and Erik Conway, *The Merchants of Doubt: How a Handful of Scientists Obscured the Truth on Issues from Tobacco Smoke to Global Warming*, Bloomsbury, 2010.

18 D.F. Ferretti et al., 'Unexpected Changes to the Global Methane Budget over the Last 2000 Years', *Science* 309:5741, 2005, pp. 1714–717.

19 'Share of renewable electricity production in OECD countries 2009–2019', International Energy Agency, iea.org. It's also worth looking at the huge advances in some non-OECD countries, including in South America and Africa. With installed renewable electricity like offshore wind now cheaper than fossil fuel produced electricity, the next decade will see a dramatic increase worldwide in renewable electricity.

20 Our World in Data, ourworldindata.org

21 Sigurd Magnusson, 'NZ Electric Car Guide 2020', Leading the Charge, leadingthecharge.org.nz/nz_electric_car_guide.

22 'Norway's electric car market has overtaken traditional vehicle sales', World Economic Forum, 5 April 2019, weforum.org/agenda/2019/04/norway-electric-car-market-vehicle-sales. Practically every month a new sales record is reached for electric cars in Norway, with current statistics showing that sales of electric cars have been the least affected of any industry by the Covid-19 pandemic and providing hope for a shift to less carbon-intensive transport.

23 Corinne Le Quéré et al., 'Temporary Reduction in Daily Global CO_2 Emissions During the Covid-19 Forced Confinement', *Nature Climate Change* 10, 2020, pp. 647–53.

24 'Covid-19 Coronavirus Pandemic Live Update' at worldometers.info/coronavirus/

INDEX

italics denotes illustrations

pic denotes photographs in the insert between pages 128 and 129

Aachen (Germany) 186–7

air pollution 13–15, 17, 24, 50, 65–6, 158–9, 248

   see also pollution

air sampling

   atmospheric CO₂ *pic*, 15–17, 56, 82–4, 117

   formaldehyde *pic*, 146–7, 150, 154–9, 162–75, 179–82

   methane 202–4, 210

airborne fraction 84, 98

Alaska 13, 234, 236–9

American Chemical Society (US) 238

American Geophysical Union 166–7, 179–80

AMS (Accelerator Mass Spectrometry) *pic*, 194–204, 206–7, 210, 259

Antarctica 87, 89–91, 157, 168–9, 218, 221

   see also South Pole Amundsen-Scott station

anti-nuclear movement 185

Anza-Borrego Desert (US) 78, 124

AR4 report (*Climate Change 2007: The Physical Science Basis*) 226–8, 235

Arctic 13, 221, 247

Ardern, Jacinda 245

Arrhenius, Svante 14

atmosphere 14, 17–18, 23–4, 66, 72, 157, 191

   damage to 25, 29, 158, 193, 201, 221–2, 234, 239, 243–7

   see also atmospheric CO₂; environment

atmospheric chemistry 97, 111, 138–40, 222, 225

   research 154, 172–3, 197, 204–7, 213–14, 216–17

   see also IGAC; Lowe, David

atmospheric chemistry institute *see* ICH3

atmospheric CO₂ 35, 39, 47, 50, 184, 191–2, 201, 239

   absorbed by oceans *see under* oceans

   extra CO₂ 14, 58, 84, 89–90, 95, 98, 119, 243–4

   measurements 11–12, 15–19, 59, 67, 76–9, 81, 95, 114, 192–3, 219, 241

   reference gases *see* calibration gases

   seasonal variations 83–4, 96, 122

   see also carbon emissions; climate change; Keeling Curve; Lowe, David; *specific sites*

atmospheric formaldehyde 140–1, 144–7, 150–1, 154–9, 162–75, 179–82, 196

atmospheric methane 157, 159, 175, 219, 221

   graphite target preparation 200–7, 210, 212–14

   photochemical modelling 172, 180

   removal by OH free radical 139–40, 154, 165, 196

aurora phenomenon 90

Australia 13, 247

Bacastow, Bob 130

Bainbridge, Arnold *pic*, 20, 39–40, 45–52, 54, 57, 70, 132

Baring Head (NZ) *pic*, 11, 17–18, 64, 72, 167

   data and article 100, 119, 121–3, 125–7, 129–35

   initial programme 60–2, 67,

264

INDEX

INDEX